AN INTRODUCTION TO

Shared Mobility

D. Philipson, S. Carroll & C. Cox

Copyright © 2022 by David Philipson, Steve Carroll and Chris Cox

All rights reserved.

No part of this book may be reproduced in any form or by any electronic or mechanical means, including information storage and retrieval systems, without written permission from the authors, except for the use of brief quotations in a book review.

All web links were accurate at the time of publishing. However, as these are external links we take no responsibility for maintaining these links.

Contents

Preface

The world is changing at a faster rate than ever before. Scientists across the globe are observing changes to the Earth's climate, and while changes to global and regional climate are not unusual, based on what we know of the planet's history, the rate, range and level of change are unprecedented in thousands, if not hundreds of thousands of years. We are now reaching the tipping point where the impact of these changes may become irreversible.

According to IPCC Working Group Co-Chair, Valérie Masson-Delmotte:

"It has been clear for decades that the Earth's climate is changing, and the role of human influence on the climate system is undisputed."

Carbon dioxide and other greenhouse gases are now confirmed to be the major contributors to climate change, and according to the IPCC's Sixth Assessment Report (AR6), emissions of greenhouse gases from human activities were responsible for around 1.1°C of warming since 1900; if nothing changes, global temperatures are expected to exceed 1.5°C of warming by about 2040.[1]

As a result, the world has been thrown into a frenzy of low carbon technology development and net zero strategies. While every area of life is being impacted by these changes, the industry most impacted by this agenda is the transport industry.

The first step in tackling this challenge has been to shift from fossil fuels to renewable energy and net zero fuels such as electricity and green hydrogen. However, converting all our existing vehicles to net zero fuels won't be enough to avoid exceeding 1.5°C of warming by about 2040. To achieve this, a much more fundamental change is needed: moving away from existing

1 Climate Change 2021, The Physical Science Basis, Summary for Policy Makers (https://www.ipcc.ch/report/ar6/wg1/downloads/report/IPCC_AR6_WGI_SPM_final.pdf)

transport solutions and towards using the right transport option for the right journey; or in simpler terms, the right tool for the job! This is the key belief which drives the shared and sustainable transport movement.

"Using the right mode for the right journey will be an important factor as we decarbonise transport. The use of shared and sustainable transport options, such as micromobility vehicles like e-bikes and scooters, offers an opportunity for us to use resources in a smarter and more efficient way whilst reducing carbon emissions."
– Beth Morley, Mobility Project Manager at Cenex

While no one handbook can cover every development in the industry, this one explains the available technologies and opportunities for shared and sustainable transport, providing a logical and relatively comprehensive guide to develop the reader from novice to expert, and signposting other important resources to help along the journey.

Chris Cox, Cenex, 2022

About the authors

David Phillipson

David is a qualified mechanical engineer. As a Technical Specialist in the Cenex Transport Strategy and Mobility team, he provides technical skills and project management to deliver customer projects related to the development and deployment of new and emerging low emission vehicle technologies.

His key focus is the delivery of strategic projects to identify and implement technological solutions and policy, to improve the environmental performance of the transport network. This includes forecasting future shared mobility deployment and modal shift, biofuel, hydrogen and electric HGV uptake in regional areas; strategic recommendations for zero emission bus deployment; and researching future trends in Connected and Autonomous Vehicles (CAVs).

Steve Carroll

Steve is an experienced project engineer with a BEng (Hons) in Manufacturing Systems Technology and an MSc in Renewable Energy System Technology. With over 15 years' experience in the automotive, power generation and renewable energy industries, Steve is Head of Transport, responsible for managing the Cenex portfolio of low carbon vehicle projects, sustainable mobility consultancy advice and vehicle demonstration trials. His project experience includes leading work on the analysis, reporting and dissemination of various sustainable mobility, hydrogen, electric vehicle and biofuel (including biomethane and biodiesel) initiatives and studies.

Chris Cox, MIMechE

After more than a decade working on research and innovation projects in the energy sector, Chris joined Cenex in 2018 to lead Cenex's activities relating to energy and infrastructure for transport; he also oversees Cenex training activities. Chris is a certified project manager, chartered engineer and member of the Institution of Mechanical Engineers (IMechE). Since taking the plunge by replacing his car with an e-bike in 2021, Chris has become a strong advocate of using the right form of transport for the right sort of journey – an ethos which is at the heart of the sustainable transport and shared mobility movement.

Wider acknowledgements

Many others at Cenex have contributed either directly or indirectly to the content of this handbook. Staff at Cenex have been working on shared mobility since 2017. Projects have included InclusivEV, SuSMo (Sustainable Urban Shared Mobility) and Nottingham City Council EV Car Club Community Engagement. Cenex staff hosted the SuSMo (Sustianable Shared Mobility) project's webinar series, which shared learnings from partner cities across Europe; this project also yielded a guidance report, written up by Cenex: 'E-Scooters: Maximising the Benefits of E-Scooter Deployments in Cities'. Cenex also worked with CoMoUK to compile the report, 'Shared Transport: An Action Kit for Employers'. Of particular mention among Cenex staff are Beth Morley, who has led many of these projects, Fergus Worthy of Cenex Scotland, as well as Theodora Skordili and Daan van Rooij from the Cenex Netherlands team.

In addition to those at Cenex, we have been fortunate to work with a number of outstanding partners whose help and support has contributed to this knowledgebase. In particular:

→ EIT Climate-KIC, funders of the SuSMo (Sustainable Urban Shared Mobility) and InclusivEV (car club) projects

→ SuSMo Project partners: Cenex Nederland, Trivector, TU Delft,

Cleantech Bulgaria, AESS-Modena and Centro de Inovacion de Infraestructuras Inteligentes

➔ InclusivEV Project partners: E-Car Club, AESS-Modena, University of Valencia UVEG - LISITT, Instituto Technologico de la Energia (ITE), Accord Group, Carplus, Paterna City Council, Wroclaw University of Economics, Lower Silesian Regional Development Agency

➔ Nottingham City Council EV Car Club Community Engagement Project partners: Enterprise Car Club and Nottingham City Council

➔ Shared mobility charity CoMoUK

➔ Cenex mobility hub stand partners at LCV 2021: Coventry City Council, Comms365, Enterprise, Ginger, Hewitt Studios LLP, RS Electric Boats, WiCET Project

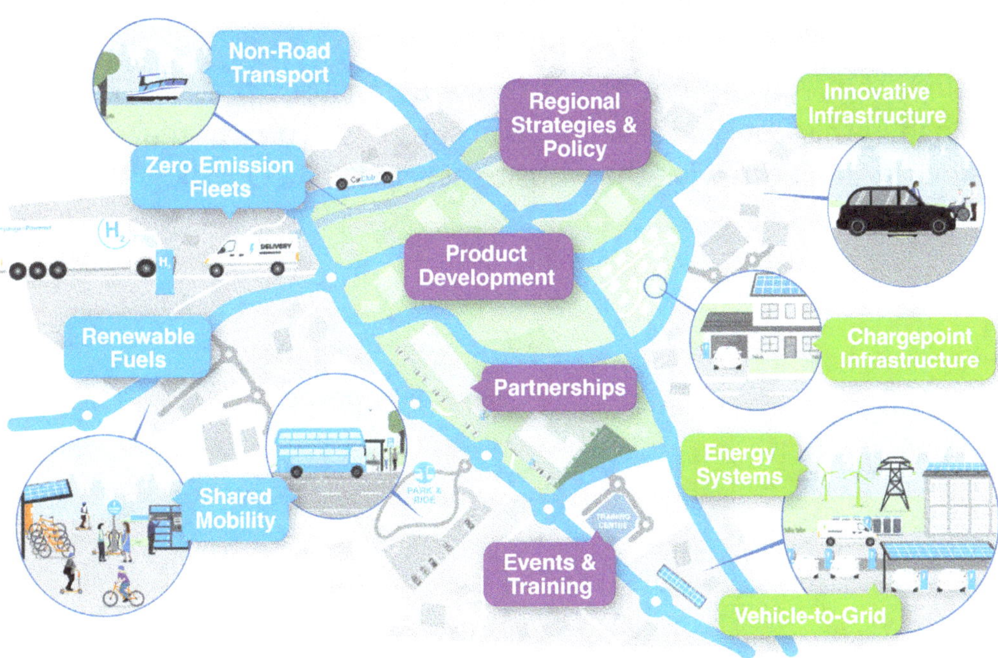

About Cenex

Cenex, the UK's first Centre of Excellence for Low Carbon and Fuel Cell Technologies, is an independent, not for profit, consultancy and Research & Technology organisation.

Transport emissions continue to rise, causing environmental issues such as climate change and poor urban air quality that affect us all today; they will continue to do so in the future unless we act now.

For the last 15 years Cenex has specialised in lowering emissions through innovation in transport and energy infrastructure; our key focus is on low emission transport, including battery electric, hydrogen fuel cell, and biofuels, as well as the innovative charging and fuelling infrastructure required to support them.

It has worked with governments and organisations across the globe to transition to net zero transport technologies, and its team of specialist experts have created this handbook so that you can develop your understanding of the history, technologies and opportunities for sustainable transport and shared mobility.

Their independent and not-for-profit status ensures that the knowledge contained in this book is reliable and accurate, so you can make informed decisions in the future. Visit the website at www.cenex.co.uk

Introduction

Learning objectives

When learning something new, it is always important to be clear about what you want to get out of it so that you can focus your learning and get the most out of your time.

Our aim is that by the time you've worked your way through this book you will be able to:

→ Evaluate what shared mobility is and how it can reduce transport emissions.

→ Compare different shared mobility modes of transport and gain an awareness of their applications.

→ Reflect upon the current limitations and benefits of each shared mobility technology and what the future may hold.

→ Discuss how shared mobility fits into the current transport model and how policy can affect the uptake and deployment of shared mobility.

→ Identify the different types of mobility hubs and how they should be deployed in order to maximise uptake of shared mobility.

→ Gain an awareness on the topic of data and data sharing and how this will play a pivotal role in the future uptake of shared mobility

→ Identify the right technologies for different applications and understand the key factors which impact this.

Of course, any good book is only helpful if it's helping you. So do start by analysing how you want to apply what you will learn to your own life, workplace and community.

Content covered

Part 1 An Introduction to Shared and Sustainable Transport

Part 1 begins by setting the scene for shared mobility and sustainable transport, and why it is likely to play a role in the future decarbonisation of road transport.

You will explore the main benefits of shared mobility, as well as what shared mobility is and what it hopes to achieve through its deployment.

Importantly, you will also discover what the challenges are to deploying shared mobility transport, both in urban and rural settings, looking at the technological and cultural barriers to mass deployment of shared mobility.

You will start your shared mobility journey with car clubs and car sharing, looking at the applications and limitations of the technology, as well as the future of this mode of transport and what path the technology is likely to take.

Part 2 What is Micromobility?

In Part 2 you will explore two more forms of shared mobility: electric scooters and electric bikes. These are both categorised as micromobility, a concept we will explore in this part. As with car clubs, you will explore these modes of transport, from their benefits and limitations to what the future may hold.

Part 3 How is this Applied to City Planning and Mobility Hubs?

In Part 3 you will explore how this knowledge can be applied in the real world. This includes where these vehicles should be deployed for maximum environmental benefit and uptake, as well as what cities can do to help support this. Finally, you will look at the concept of mobility hubs and how data is changing the ways of the world and enabling the shared mobility industry to grow.

Who is this book for?

This book is targeted at both people early in their career and those looking to transition from existing employment into the low carbon transport economy.

It is designed for anyone interested in the ongoing change in the shared mobility sector, which may include those who work in transport planning, sustainability or even social equality. It will particularly benefit you if you have a basic understanding of transport technologies.

The net zero transition has a significant impact on many people already in employment, with rapidly rising demand for skill sets which are currently relatively niche and far exceeding the pipeline of graduates and apprentices. As a result, the low-carbon transport industry is dangerously short of experienced workers, especially those in the PEMD sector, which if left untackled will impact upon the growth of the industry. This means that while the net zero economy grows, it is essential that there is support both for those early in their careers and for those looking to retrain.

The oil and gas industry alone employs around 150,000 people worldwide. In the UK alone, over 30,000 people in the UK are directly employed each year by the oil and gas industry, and on average a further 150,000 people were indirectly employed by the industry. Similarly, according to the SMMT, the automotive industry directly employs around 180,000 people in manufacturing in the UK, and in excess of 864,000 across the wider automotive industry (including suppliers).

The transition to net zero, shared and sustainable transport technologies will have a significant impact on both the oil/gas and automotive sectors, with many people finding that they do not have the understanding or awareness of what is involved in low-carbon transport to be able to make the transition easily into new roles as their existing roles become redundant.

This book therefore offers a stepping stone, for learners of all levels, to gain a fuller understanding of sustainable transport and shared mobility.

How to use this book

Each section will make use of the knowledge built during the previous sections(s), so it is helpful, though not essential, to go through this book in order. However, you may find that certain sections are more relevant to your job or interests. To help this, as much as is practical, each chapter has been written to make sense on its own.

Text in bold prompts you to question and apply what you're reading to your own circumstances.

At the end of each part there is a short quiz aimed at testing how much you've taken in. The answers to these quizzes can be found at the end of the book. It's worth having a pen, highlighter and paper with you to take notes of anything which stands out – or do highlight and make notes straight in this handbook.

Towards the end of the book, there is a 10-question final test which draws from content from all three parts. This will allow you to assess how well you have retainedyour knowledge.

As the world of low carbon technology is filled with acronyms, and there isn't always clear consensus on their use, this book also contains a glossary of terms, which can be used as a reference guide as you work through the sections. This is also found towards the end of the book.

Throughout this book we have attempted to display the key content to make it easier to refer back to. In particular we have drawn out *quotes*, **definitions** (with the term in bold, followed by the definition), and **key takeaways**.

AN INTRODUCTION TO SHARED AND SUSTAINABLE TRANSPORT

Part 1 sets the scene for shared mobility and sustainable transport and why it is likely to play a role in the future decarbonisation of road transport. The following sections will explore the main benefits of shared mobility, as well as what shared mobility is and what it hopes to achieve through its deployment.They will also demonstrate the challenges to deploying shared mobility transport, both in urban and rural settings – including both techno-logical and cultural barriers.

1.1 What is shared mobility?

Shared mobility is a term that refers to modes of transport where an operator owns a vehicle and the public can rent it over a period of time for their own personal use.

Users share a vehicle over time as a personal rental when they require it, and they rent the full vehicle; this is where it differs from public transport. Shared mobility removes the need for the user to purchase a vehicle, and instead the operator owns the vehicle and allows others to rent it for a period of time. In the process, the users share the cost associated with owning the vehicle, creating a hybrid between private vehicle use and mass or public transport.

Shared mobility can reduce the carbon footprint of transport, moving people from carbon intensive cars to more sustainable forms of travel. Though electric cars are less carbon intensive than petrol and diesel vehicles, shared mobility can reduce this impact further, emitting less carbon dioxide per mile travelled. However, the use of cars in shared mobility can still lower overall emissions, with newer, less polluting vehicles available to rent that would otherwise be outside the price range of users.

Shared mobility allows users to access transportation services on a 'need-to-use' basis, removing the need for large capital investment costs. This allows the general public to access sustainable travel regardless of socio-eco-nomic status. Developments in apps and the use of smart phones allow easy and quick access to these transport modes. In the future Mobility as a Service (MaaS) apps will provide a single platform for all shared mobility modes, allowing a one-stop-shop for users, instead of requiring multiple travel-based apps for each operator.There are a growing number of shared mobility services provided by operators, as this market begins to develop. The key shared mobility modes that this book will cover are as follows:

→ Car clubs and car sharing: This differs from a conventional rental scheme, with no restrictions on where and when you can rent; a whole fleet of different vehicles are available to you at the tap of a button

→ E-scooters: A simple to use new form of shared mobility, providing a platform to stand on and full battery power, requiring no effort from the user

→ E-bikes: An electric assisted version of a standard bicycle, allowing you to travel further, and with greater ease, than a conventional bike

---

KEY DEFINITION:

Shared mobility - modes of transport where an operator owns a vehicle and the public can rent them over a period of time for their own personal use.

Mobility as a Service (MaaS) - A digital service which allows users to plan, book and pay for multiple shared transport services through one place.

1.1.1 The main benefits of shared mobility

Now that you understand thebasic shared mobility definitions, let's look in depth at the associated benefits.

Sustainability

For all mobility solutions, sustainability and the environment must be at the forefront of decision-making.

Both technology providers and operators should commit to sustainability and minimise the environmental impact of the services provided. This could be through sustainable methods of producing the vehicles and ensuring components can be recycled at the end of life, or through how vehicles are maintained to extend the usable life of these vehicles.

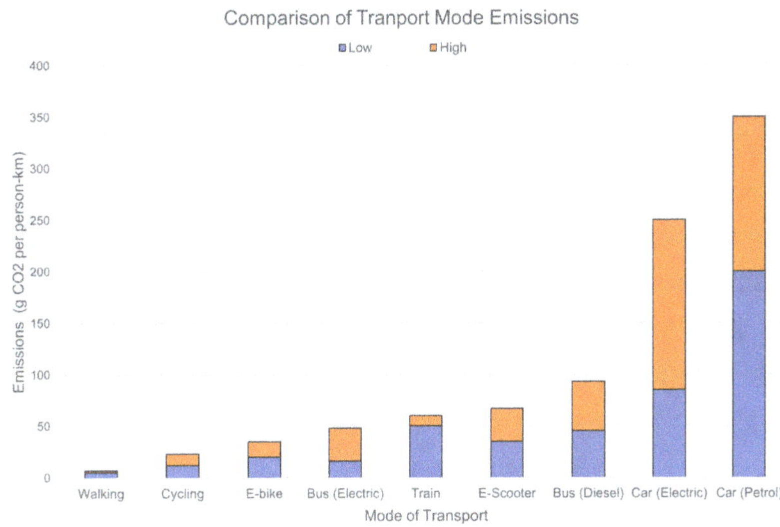

Figure 1 Transport mode comparison: range of gram CO2 per person-km

Providing a sustainable mobility service that deploys quiet electric vehicles, and promotes a shift away from private car use, can help reduce greenhouse gas emissions (GHG) and improve air quality.

It is well known that electric cars are better for the environment than conventional petrol and diesel cars. However, when comparing GHG emissions to other modes of transport, there are more sustainable options, including shared mobility.

Figure 1 shows typical ranges of carbon dioxide emissions for a variety of transport options. Here it can be seen that shared mobility options, along with public transport, can achieve lower emissions than cars (even electric ones).

Accessibility and ease of use

Shared mobility can benefit citizens of every socio-economic status. Unlike passenger cars which require large amounts of capital investment or expensive lease schemes, shared mobility has no capital outlay for the end user. The user will just pay for the journey they need. Typical user costs for some of the main forms of shared mobility are stated below:

➜ Car clubs – Monthly membership fee from £7/month, Hourly rental rate from £5/hour

➜ Electric scooters – From £2 for a 2-3 mile trip

➜ Electric bike – From £2 for 30 minutes

➜ Non-electric bike – From £2 per day or annual memberships from £90+

Shared mobility can also improve transport access in areas where public transport is scarce. These gaps in the transport network can be plugged by shared mobility for a low cost. Through the integration of shared mobility into the transport system, this will give citizens greater accessibility to cheap transport.

Reduced congestion and journey times

Congestion represents wasted time for car drivers, with an average of 178 lost hours per year in the UK and an associated cost of £8 billion. Lost time due to congestion is most common in city centres where roads are often single lane and there is a high ratio of traffic crossings to road space. Additionally, limited parking in city centres can extend journey times, through drivers having to search for available parking spaces. Due to the urban environment, expanding the road network is difficult. Often the only way to reduce congestion is to reduce the number of vehicles travelling through the city.

Figure 2 Vehicles stuck in traffic

Shared micromobility (e-scooters, e-bikes and conventional pedal bikes) can provide a solution for this. These smaller forms of shared mobility have higher manoeuvrability through traffic, as well as users being able to use cycle lanes, where available. Typical speeds for electric scooters and electric bikes are between 12.5 mph and 20 mph (20-32 kph). Given typical inner-city speeds of around 10 mph (16kph), shorter journeys can be carried out more quickly using shared mobility and can help to reduce congestion in cities as people choose shared mobility over using their own car.

Additionally, some forms of shared micromobility are dockless. Dockless vehicles can be parked anywhere within a pre-specified zone, allowing users to travel from door to door. This can further reduce journey time, removing the need to try to find somewhere to park and then walking to your final destination.

Research into car clubs shows that they can reduce the number of vehicles on the road, with one vehicle replacing between 8 and 12 privately owned vehicles. The spread of car clubs can therefore help reduce the amount of space that privately owned vehicles take up within a city, easing the pressure on parking as well as congestion.

The benefits of shared mobility are many and can help improve travel, decrease cost and reduce the impact of transport on the environment. So why is shared mobility not as prevalent as it should be? In the next section you will look at some of the challenges associated with shared mobility.

1.1.2 The main challenges of shared mobility

Now that you have considered the benefits of shared mobility, it is time to look at some of the challenges

Cooperation with cities

Operators of shared mobility and the local authorities where they are deployed may have conflicting views on how best to serve the public. As is often the case with new technologies, shared mobility vehicles came before any specific policy or legislation. With no regulation or transport policy, some cities suffered from large-scale deployment of the smaller forms of shared mobility (e-scooters and e-bikes).

Citiy authorities are starting to catch up with these new shared mobility technologies and beginning to understand the benefits. However, it is key if shared mobility is to succeed that cooperation between operators and cities is in place from the start. This includes working with each other, and the community, to improve the sustainability of a city's transport system.

Public transport and active travel could form the backbone of sustainable urban mobility. Shared mobility services could then support these modes rather than compete against them. This should be communicated to operators, so they understand the aims of a city's transport network. This will ensure that the operator can contribute to the city's sustainability objectives, and not just their business case.

Space

Cities have historically allocated a large amount of public space to privately owned vehicles. This bias towards private cars reduces the potential for deployment of more sustainable shared modes of travel with limited space in which to deploy vehicles, and limited space to use them in a safe environment (e.g. cycle lanes).

The reallocation of space for the safe use of active travel and shared mobility is not about penalising private vehicle use; it is about creating a network that allows more sustainable modes of travel to be a genuine option for those who want to use them, whilst feeling safe and seen by other road users.

The proliferation of e-scooters and e-bikes has also paved the way for cities to unlock their public space for future innovative modes of transport. This will enable citizens to become less car dependent, as it becomes easier to select and use more sustainable ways of travelling.

Infrastructure

Provision of suitable parking for shared mobility is important to encouraging an environment that promotes these forms of transport: whether that is car parking spaces and charging infrastructure for car clubs, or areas to park e-scooters and e-bikes that will not impede pedestrians on footpaths. Howeverspace is in short supply in city centres.

It is important that locations provide good visibility of the service: a hidden away car park with little footfall is not a good location for promoting the use of shared mobility.

Reallocating public parking spaces can be fast and cheap, with only signage, road paint and barriers needed. One car parking space can be replaced with another for a car club vehicle, or allow up to 5 e-bikes or 20 e-scooters to be parked. Additionally, the decreased parking allocation for privately owned cars may increase demand for shared mobility. These strategies, though beneficial, may see resistance from citizens if communication is poor.

Safety

It is important with any strategy for shared mobility that the general public are well informed. New technologies can be overwhelming when they first arrive, and it is important that citizens understand them and their purpose. Only when the general public accept these new forms of travel will the challenges of deploying shared mobility be mitigated.

For the smaller forms of shared mobility (e-scooters, shared bikes and e-bikes) there are three main safety challenges posed: that of the user, pedestrians, and traffic safety. It is important to look at possible safety solutions for users, such as helmets, high visibility vests and reflective bands, though these can be difficult to enforce.

However, often the greatest threat to the safety of riders is the danger from other road users, beyond the scope of protective equipment. Many of the high-profile incidents and deaths from people using e-scooters and bikes have been caused by accidents with other road users (cars, vans, buses and heavy good vehicles). These incidents can have a negative impact on the public's perception of shared mobility being a safe mode of travel, even if

Figure 3 A two-way cycle path; segregated cycle paths can provide a safe way for shared mobility users to travel away from busy roads

the user is innocent and the encounter out of their control. Reducing the speed of traffic, introducing cycle lanes and traffic calming measures can at least enhance the safety of shared mobility users and reduce the number of incidents and deaths associated with these vehicles.

Limitations can be placed on users of shared mobility, such as requiring a driver's licence or setting age limits for users. Although this can make sense from a liability angle, there is a trade-off with the number of people able to use these services, since shared mobility is a popular mode of transport for younger generations. While age limits and licence requirements can protect operators and cities from liability, they are not necessarily enough to protect individuals from potential harm.

Safety for pedestrians links inherently to infrastructure provision and where vehicles are parked. People with visual impairments need a functional walking path without unexpected obstacles. Random dispersion of parked scooters and bikes can create trip hazards and force frequent changes in direction. Providing suitable infrastructure and guidance on parking is needed to ensure clear and safe footpaths.

While there are many challenges to overcome, there are usually clear and achievable solutions to these problems.

For the remainder of Part 1, we will look at car clubs and how this kind of shared mobility works.

1.2 Car clubs

1.2.1 How do car clubs work?

Car clubs, known as car sharing in mainland Europe, can be thought of as a short-term car rental that gives you the convenience and benefits of vehicle ownership without the costs and responsibilities.

Car clubs are a membership-based service that provides its members with access to a variety of vehicles. For a small membership fee, users can reserve vehicles suited to their needs. This membership can remove the need to own a personal vehicle, which can be expensive.

Vehicles can be reserved in advance or on the go, allowing flexibility for members. Once a vehicle has been selected, an app on your phone can be used to unlock the vehicle, typically followed by a PIN number to access the keys inside the vehicles. During the hire period, the keys will operate like any normal vehicle, allowing you to lock and unlock the vehicle. Once you have finished with the vehicle users can drop it off (at a designated location), return the keys to the safe location inside the car, and lock the vehicle using an app on your phone.

There are a number of different operating models used and their popularity varies depending on the country. The two main operating models used are station-based, and free-floating.

Station-based

Station-based car club organisations are most similar to traditional car rental companies. A vehicle will have a designated location from which it can be rented and then returned. There is increased flexibility compared to traditional car renting, in that vehicles are rented hourly and can be rented at any time, day or night. Additionally, no advanced booking is required. Insurance and eligibility to drive are completed when you join as a member, making the process quick and simple. In the case of the Enterprise Car Club

Figure 4 Enterprise Car Club vehicle: charging in a dedicated parking spot

(Car Share in North America), once you are a member, you gain 24/7 access to any vehicle with no prior booking required.

Free-floating

Free-floating car clubs make vehicles available for users within a designated area. Users of a free-floating car club can book a vehicle and complete a trip to a second location within the designated 'home zone' of the car sharing service. This model allows flexibility of use, with users having no requirement to return the vehicle to its pick-up location when finished.

An issue that may arise is vehicle concentration or scarcity within the designated area, sometimes referred to as the 'relocation problem'. The relocation problem requires a system by which either additional employees are paid to relocate vehicles or incentives are used to motivate users to park within a specific area. In the future, shared autonomous vehicles (SAVs) could solve the relocation problem, making car sharing more viable. Despite the relocation problem, a free-floating car club can avoid the infrastructure costs that station-based models require, such as owning land for a depot.

In addition to serving the general public, car clubs can also serve business needs. For small and medium businesses (SMEs), purchasing vehicles outright or on lease may not be possible due to high costs versus relatively small turnovers. Car clubs can provide customised plans to allow organisations to sign up to their vehicles. This can help in reducing travel expenses as well as helping to meet sustainability goals. Often car clubs will offer reduced rates and fees via their business specific plans.

Case study: EV Car Club in Nottingham, UK

Nottingham City Council and Enterprise Car Club have an ambition to expand car club locations into lower-income neighbourhoods traditionally not serviced by new shared mobility modes.

The aim of the project was to test whether deploying an electric car club vehicle without a dedicated parking bay or chargepoint was a viable solution for the car club provider, potential users and the local authority.

The project applied behaviour change techniques to help the local authority and the car club provider understand the complex barriers and incentives that led to modal shift, and how these elements should be considered when planning and implementing a low carbon transport solution.

By working with influential representatives to foster ownership and anchor the car club into the community's culture, Cenex increased interest in, and support for, the car club throughout the community. The first car location was launched and showed excellent usage in the first few weeks. In the second location, knowledge of the car club grew and some potential local champions have been identified. Cenex facilitated the development of strong relationships between the stakeholders.

The project demonstrates the value of using community engagement as a tool for promoting modal shift and supporting alternative models of deployment for EV car clubs.

KEY DEFINITION:

Car clubs – a short-term car rental that facilitates the convenience and benefits of vehicle ownership without the costs and responsibilities of ownership.

1.2.2 Applications, benefits and limitations

Applications

Car sharing schemes require higher population densities; good pedestrian and cycling environments; good public transport (for transport to and from car club locations); and the availability of car parking spaces. These metrics should be analysed to ascertain if the selected location is suitable for a car club, or if other changes are required first.

If these metrics are all met, then defining the strategic vision for the car club is vital to its success. Is the aim to increase the mobility and accessibility of low-income areas, to reduce congestion, to reduce emissions, or perhaps all three and more? Understanding what the scheme is attempting to achieve will be key, prior to setting out an implementation plan or measures of success.

For example, the case study described for Nottingham had the priority objective of increasing accessibility for low-income areas. This clear vision was at the heart of the strategy for implementation, and ensured that locations were appropriate for its aim, and that promotion was directed towards the key demographic that the car club was targeting.

Social benefits

Fairness and equality are often described as some of the benefits of car clubs, with lower-income individuals gaining greater access to transportation options where car sharing schemes have been implemented.

This may be important, as there is literature that indicates that having access to a car is positively correlated with improved job opportunities. A 2019 UK Government Office for Science report also highlights the inequalities in mobility that follow from public transport limitations, resulting in less access to employment, education, healthcare, shops and services. For those low-income families that may feel forced into vehicle ownership to address these disadvantages, this can lead to economic stress due to the high cost of owning and maintaining a vehicle.

Community and environmental benefits

There are many likely benefits to the community from increased car sharing. These include less congestion, reduced parking pressure, less energy/resources expended for vehicle manufacturing, and a reduction in pollutants and greenhouse gas emissions.

These benefits are reliant on a reduction in vehicle usage, which matches the findings of multiple literature sources, such as a 2010 report on the state of European car sharing which found that each car in a car sharing scheme could replace up to 8 personally owned cars, as well as reducing average annual mileages for car club members. Other studies by organisations such as CoMoUK state that one car club vehicle may even replace up to 12 personal cars. The reduced vehicle mileage per person is because if members do not have a personal vehicle available to them, they are more likely to use other forms of transport as an alternative, including public transport and active travel.

Figure 5 Car club vehicle from the InclusivEV project

In addition to reducing the number of vehicles on the road, there is also the potential for the car share vehicles to have an improved fuel economy, if the car club vehicles are newer and more advanced than personally owned vehicles, which may result in greenhouse gas reductions. Whether this reduction in on-road vehicles results in emissions reductions may be a contentious issue. Some literature indicates that car sharing may reduce greenhouse gas emissions by 0.1-2.6%. Conversely, some reports indicate that any offset in greenhouse gas emissions due to a move from individual ownership to car clubs is offset by the shift of many journeys from public transport to car shares.

An additional benefit of car clubs is an ability to switch to low emission vehicles more quickly, due to the faster turnover rate of shared vehicles and the buying power of operators. This could facilitate a faster paced transition to electric vehicles, with a consequent reduction in emissions.

Economic benefits

A 2019 publication by the UK government found that low-income households spend up to 40% of their income on motoring costs. In Europe, a 2017 EU report found that the average cost of car ownership was 6,500 euros per year.

Car clubs remove the capital investment of ownership and can therefore reduce the overall household expenditure on transport. Also, as fuel costs, insurance, initial purchase price or finance charges are ownership costs that leave the local economy, car sharing may boost spending within the local community due to members having more spare capital.

Limitations

While car clubs may reduce the overall environmental impact of road transport, the lifecycle emissions of cars, electric or not, are more polluting than other forms of transport. While they do provide some benefits, public transport and active travel should still be seen as preferred forms of transport if appropriate for the journey.

Car clubs may still exclude those with a physical or cognitive disability that means they are unable to operate a vehicle. Additionally, an inability to afford the cost of hiring the vehicle, or any difficulties for car sharing schemes operating in specific areas, may limit the benefits of car sharing in regard to reducing inequalities.

One specific prejudice could be that areas with high crime rates increase the likelihood of vehicles being vandalised. During the InclusivEV trial in Solihull, a high level of vandalism occurred and forced the trial to end 3 months into an 18-month programme. A spokesperson for the car club said, 'Due to the exceptional levels of damage to the vehicles caused by vandalism, E-Car were forced to remove the vehicles and suspend operation of the Solihull scheme in February 2019.'

Space may also be at a premium in inner-city, low-income neighbour-hoods, and finding somewhere suitable to place these vehicles may prove challenging. Finally, depending on how the car club scheme is accessed, over-reliance on apps and technology could exclude those with no access to a mobile device.

For car clubs to realise their maximum potential fairness and equality benefits, the factors above will all need to be considered.

1.2.3 Car club roadmap

It is important to understand the future landscape of the transport network in order to inform the decisions made today. Though forecasting technology developments cannot be an exact science, there are clear indicators today that help inform the likely trends of future shared mobility transport.

Through understanding the timeline for technology advances and changes in operation, cities can ensure that their policies and strategies are sustainable for the long term.

The roadmap in Figure 6 displays an indication of the likely route that car clubs may take through to 2030, with the text below explaining the likely reasons for these changes. These roadmaps were developed through the Climate-KIC funded project SuSMo, investigating sustainable and shared mobility throughout Europe, creating tools for cities to accelerate the uptake of shared mobility.

Technology

The future technological development of car clubs will focus on two main areas: electrification and autonomous vehicles.

The electrification of car clubs will be driven by the operational cost savings that electric vehicles (EVs) have over internal combustion engine (ICE) vehicles. Cost parity for new vehicles will be reached by around 2025. However, it should be noted that over the life of a vehicle, EVs are already cheaper than ICEs due to their lower fuel and maintenance costs. Beyond 2025, the increased availability of EVs and yet more cost reductions due to economies of scale will likely accelerate EV adoption within car clubs. This adoption will be helped by municipalities adopting popular strategies to combat climate change, such as ultra-low emissions zones and more demanding vehicle emissions standards.

Prior to 2025, the level of autonomy seen in vehicles will be relatively low and limited to technologies such as lane-assist and automated cruise control.

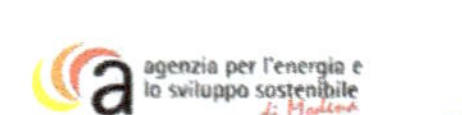

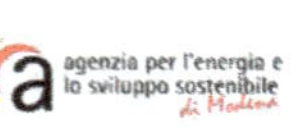

Figure 6 Roadmap for car clubs, showing possible future developments through to 2030

Towards the end of the decade, as autonomous technology continues to develop, early adopters of autonomous technologies are likely to deploy vehicles within urban environments as part of a car club or taxi fleet. The economic case for removing a driver, and the ability of autonomous vehicles to solve the relocation problem within certain car share models, make its use highly desirable for car clubs.

Operations

Car club models are likely to be a mixture of free-floating and station-based. They will vary on a local basis, with free-floating most popular in Germany, and a slight preference for station-based in the UK, although both models will be popular. The electrification of car clubs' fleets will slow the growth of the free-floating model, in preference for station-based. However, as charging infrastructure spreads and autonomy progresses, free-floating is likely to become the dominant model.

From 2025 onwards Mobility as a Service (MaaS) systems may begin to appear throughout UK and EU cities, and car clubs will be a part of this. This will help to link multiple forms of transport up into one simple payment per journey through a single app, or ticket. One of the primary advances will be a change in the management of data and its exploitation to improve the transport systems of cities.

Environment

Several factors will put pressure on car clubs to reduce the emissions of vehicles selected for use within their fleets: these being low emissions zones, congestion zones, financial incentives to choose low emission vehicles and preferential initiatives, such as parking for zero emission vehicles. Additionally, the operational benefits of using EVs, alongside the high utilisation rate of car club vehicles, will result in car clubs having lower emissions per kilometre than individually owned car trips.

As the EV industry progresses, improvements in manufacturing and recycling will see intrinsic emissions reduced, whilst the decarbonising of energy grids will see continued operational reductions in emissions.

Coverage

Projections for ridership numbers vary wildly, based on a number of drivers or barriers that could advance or hinder car clubs. In 2021, the shared mobility charity CoMoUK reported that there were more than 630,000 car club members in the UK – twice the figure from 2018 (pre-pandemic). In 2022 it was estimated that there were around 30 million car club users worldwide and, based on our own forecasts we expect to see a steady rate of growth, mainly in urban environments, seeing global car club users reaching 50 million by 2025/26 and if this growth rate is maintained, 100 million users by 2030. However, this steady rate of growth assumes that societal and regulatory barriers limit the acceptance of autonomous vehicles, and assume that no change in societal norms occur, such as any long-term impacts from Covid-19 that may limit shared vehicle acceptance.

Economics

Car clubs in certain locations today can be cheaper than car ownership. In London, for example, an individual who travels less than 2,000 miles per year will save money by avoiding car ownership and joining a car club instead. However, 2,000 miles per year is much less than the annual average of 10,000 miles. Until 2024, it is expected that car clubs will only be cheaper or comparable to ownership in the event that an individual travels less than the average national mileage per year. However, by the middle of the decade it is likely that an average user (travelling 10,000 miles) may also see annual cost savings as operator costs reduce, including from the reduced capital cost of EVs, and this is passed on to the user.

1.3 Part 1 Quiz: Test what you've learned

Well done, you've reached the end of Part 1!

You have covered a lot of ground so far, focussing on what shared mobility is, the potential benefits and challenges, and car clubs as a technology. While car clubs certainly have their benefits, as we have seen, even electric cars can produce far higher emissions per mile than public transport.

There have been a lot of new concepts to take on board, so take the opportunity to pause and see what you have learned so far.

It can also be helpful to make a note of your key learnings. Here are a few to get you started:

KEY TAKEAWAYS

1. Shared mobility allows users to access transport services on a 'need-to-use' basis, removing the need for large capital investment in vehicles. This allows the general public to access sustainable travel regardless of socio-economic status.

2. Cities have historically allocated a large amount of public space to privately owned vehicles. The reallocation of space for the safe use of active travel and shared mobility is not about fighting private car use. It is about correcting a system biased towards the car and limits the ability of those who do own a car to change to more sustainable modes.

3. Car clubs are a membership-based service that provide their members with access to a variety of vehicles. They can be deployed in two distinct ways: station-based – where a vehicle can be rented and then returned to specific rental stations; and free-floating – where vehicles are rented and returned within designated areas.

In the next part, you will explore micromobility, a smaller form of shared mobility than car clubs, which are often single-person short-journey transport options. These are often known as first and last mile, where the intention is to use micromobility at the start or end of your trip.

But before you carry on, have a go at this short, multiple-choice quiz:

1. **Which of these vehicles has the lowest carbon footprint per person, per kilometre travelled?**

    1. Electric car

    2. E-bike

    3. Diesel bus

    4. E-scooter

2. Congestion represents wasted time for drivers. What is the average amount of lost time per year in the UK due to congestion per person?

1. 17 minutes

2. 17 hours

3. 178 hours

4. 1,780 hours

3. Which of the following is not a safety measure that can be implemented for shared mobility?

1. The use of green electricity to recharge vehicles

2. Setting an age limit for users

3. Creating infrastructure for segregated parking

4. Reducing the speed of traffic

4. What are the two main operating models used for car clubs?

1. EVSE and V2G

2. Station-based and EVSE

3. Free- floating and V2G

4. Station-based and free- floating

5. If steady growth of car club members is maintained, how many global users will there be by 2030?

1. 10 million

2. 20 million

3. 50 million

4. 100 million

You can find the answers to this quiz at the end of the book.

WHAT IS MICROMOBILITY?

"You are one ride away from a good mood."
– Sarah Bentley, British Cyclist and Tennis Player

In Part 2 you will explore two more forms of shared mobility: electric scooters and electric bikes. These are both categorised as micromobility, a concept we will explore in the next section. As with car clubs, we will explore these modes of transport through their benefits and limitations to what the future holds.

2.1 Defining micromobility

Micromobility is a term associated with a rapidly evolving range of light vehicles that are now populating streets across the globe. Though there is no standard definition for what constitutes micromobility, a few common themes are recognised: light vehicles typically below 300kg, speeds limited to 15-30 mph (24-48 kph), licence requirements to operate being typically less than a car.

Below are some examples of micromobility vehicles:

➔ Pedal bikes

➔ Electric bikes

➔ Electric scooters

➔ Electric skateboards

➔ Electric hoverboards

➔ Cargo bikes

In some instances, mopeds and mobility scooters are also classified as micromobility, although there is still debate as to whether micromobility should be defined by speed or weight, since these two vehicles are both close to the boundary definitions of these categories.

Another definition sometimes used for micromobility is based on the policy of a given country as to what vehicles are permitted to use cycle lanes; any that legally can are considered to come under micromobility.

As can be seen from the list above, micromobility is not just limited to electrically-assisted vehicles; it also includes human-powered vehicles. In this book, you will focus on two of the newest and most common forms of micromobility: e-scooters and e-bikes, though the learnings can be applied to similar modes.

Micromobility has a low environmental impact, with little noise and zero tailpipe emissions. These vehicles' light weight suggests a smaller carbon footprint over the vehicle lifecycle when compared to other vehicle types (as was mentioned above). Smaller vehicles also consume less of the city's most valuable resource: space. By moving people from cars to micromobility, congestion can be eased and journey times reduced. For all these reasons, micromobility is becoming increasingly attractive to individuals and policy makers alike.

Safety

There are three distinct challenges posed by the safe operation of shared micromobility: that of the user, pedestrians and street traffic. Though it is important to look at possible safety solutions for users themselves, such as helmets, high visibility vests and reflective bands, the greatest threat to the safety of riders comes from other road users, and outweighs these gadgets.

Actions such as reducing speed of traffic, creating micromobility lanes, and introducing traffic calming measures have a much higher impact and are more far-reaching. By creating a safe environment on the road for users of micromobility, this can ensure that these modes of transport are not ridden on footpaths and walkways, which presents a danger to pedestrians. Similarly, enforcing that riders use the road, and creating geo-fenced pedestrian-only zones, can help keep pedestrians safe when shared mobility is present in a city.

The quality of the road surface is also an issue for users of micromobility. E-scooters typically have smaller, or less robust, wheels than classic forms of mobility, meaning that cracks and potholes and uneven surfaces are serious hazards to users as they ride over them, causing instability. These are much more prevalent at the side of roads and existing cycle lanes, where micromobility operates. Improving road surfaces in these areas is a quick and rational way of improving safety for users, thereby encouraging them to use the road system and thus reduce the risk to pedestrians.

Public education and awareness of micromobility is increasing constantly, as citizens become more aware of its presence. Educating citizens on how to use micromobility safely, and how to drive on the same roads and interact with micromobility, should be seen as a complementary measure rather than a substitute for other safety measures. Hundreds of thousands of people use roads in any one city, and education and awareness are unlikely to reach all users of urban environments; therefore there is inevitably a limit to the effectiveness of these programmes.

A disruptive technology?

Due to the speed of development of new micromobility vehicles, these new technologies are often launched on to the market before any specific policy or legislation is created. This often causes friction between cities and vehicle operators.

When e-scooters first entered the market, there was no regulation or transport policy, and some cities suffered from large-scale deployment of e-scooters. In some cases, competing companies distributed units in high footfall areas, with little concern for potential impacts on the city. Users of the scheme had little knowledge of how e-scooters should be used, or best practice for where they should be ridden or where they could be left. This often led to safety concerns with other road users and pedestrians, and e-scooters were viewed by some critics as potentially dangerous.

Citiy authorities and regulators are beginning to understand just how fast and dynamic the micromobility industry is, and are less likely to be caught out again when the next form of micromobility hits the market. However, the future is unpredictable, and the past quickly forgotten.

2.2 E-scooters

2.2.1 How do they work?

Shared e-scooters are a form of micromobility that has grown rapidly since their introduction in 2018, pioneered by Lime and Bird in the US. They are two-wheeled scooters with small, electric motors and a platform to stand on while riding. They are usually spread around a city for the general public to use. Companies in this field have seen dramatic market growth and huge valuations: Bird, for example, became the fastest US company to reach a $1 billion valuation.[2] Established shared mobility companies such as Uber are among the biggest investors in e-scooter start-ups, investing in both Lime and Voi.

E-scooters are predominantly dockless and users unlock the scooter through an app at a cost (around 1 £/€) plus a charge per minute (around 0.15 £/€) until they reach their destination. In some cases, they may also be required to pay a monthly or annual membership fee.

Dockless systems give users the convenience and freedom to start and end their journeys at any location, effectively traveling from door to door. However, this brings with it complications in logistics for the operators and makes vandalism and theft more common. A common approach to help mitigate this is for e-scooter providers to create a zone of operation, whereby if a vehicle travels outside that zone, it can no longer be operated.

E-scooters are intended as a sustainable method of 'first and last mile' transport, primarily targeting large towns and cities. The UK National Travel Survey of 2018 showed that 58% of car trips are fewer than 5 miles (8 km) and therefore could potentially be replaced by e-scooters, with a typical range of 40 miles (64 km). However, e-scooters typically compete with docked and dockless electric bikes and pedal bikes, and therefore may not help cities alleviate congestion. Safety is often a concern, along with street clutter, and in many places e-scooters have arrived before the much-needed infrastructure, policy and awareness to support them.

---

2 https://www.bizjournals.com/sanjose/news/2018/06/15/scooter-startups-break-unicorn-speed-records-to-1.html

Figure 7 E-scooters in Sofia, Bulgaria

The common operating model for providers is to place e-scooters in central locations close to city amenities, other transportation hubs (such as train stations), or in high footfall areas. Throughout the day their power charge decreases through use, and operators commission third-party gig-workers[3] to collect e-scooters that are low on charge in the evening and transport them to a facility where they can be recharged overnight. Once charged, vehicles are placed back in designated areas and the cycle starts again. Some operators employ swappable battery technologies, which remove the need for gig-workers to transport scooters in vans; instead they use other forms of low carbon transport, such as cargo bikes, to travel between each e-scooter. Any deployment strategy should consider indirect carbon emissions associated with e-scooter schemes when planning or procuring services.

Like many forms of transport, e-scooters have positive and negative effects on cities. It is crucial that authorities understand these in order to limit the

---

3 Gig-workers are independent contractors or freelancers who enter into formal agreements with on-demand companies to provide services to clients.

negative impacts and ensure this new form of mobility works for citizens, first and foremost, and the operators.

Case study: E-scooters in Paris, France

Lime was the first operator to offer services in Paris, launching in June 2018. During the following 12 months, the market grew rapidly with 12 operators in the city deploying 20,000 vehicles in total. This was a major problem for the city, as there was no regulation in place and the streets became littered with vehicles, considered by many as an eyesore, along with an increasing number of accidents involving both pedestrians and other road users.

This led to the creation of a framework of regulations for e-scooter companies and users by the city, to try to improve safety and control the deployment of vehicles. This included:

➜ Fines for users riding e-scooters on the pavement – €135 – and for parking on the pavement – €35.

➜ Implementation of 2,500 parking spots dedicated to e-scooters.

➜ Obligation for the operators to sign a good conduct chart.

➜ Vehicle licence fee in order to regulate the fleet size – €50/scooter/year.

However, these regulations were insufficient, as the city still had relatively little ability to monitor the service quality and compliance to the local rulings. So Paris decided to tender for operators to run services in the city.

The tender offered three companies the right to operate up to 5,000 units each within the city's boundaries, and the applications were judged according to three criteria: User Safety (30% weighting), Operations – management, maintenance, and charging (30% weighting), and Environmental Responsibility (40% weighting). Through this tender, Paris selected the best operators for their city, providing a framework for sustainable mobility.

E-scooters are a relatively new form of transport. If you have used them already, what has been your experience? And if not, why not?

2.2.1 Applications and benefits

Applications

Shared e-scooters are primarily a first and last mile mobility service; they can be used to link up other transport services, to help create a fully integrated transport network. They require no training to use, and as they don't use docking, can transport users to any final destination.

E-scooters are ideal for congested city centres, due to their compact design and door-to-door operability. Manoeuvrability through traffic means that short journeys can be much quicker than private car travel in city centres, with the additional ability of being able to park e-scooters at any destination.

Figure 8 Using phones to unlock parked electric scooters in Sofia, Bulgaria

Economic and social benefits

Shared e-scooters can benefit citizens of any socio-economic status. Unlike passenger cars, where significant capital investment is needed or expensive lease schemes, shared e-scooters can cost as little as £2/€2 for a 2-mile trip with no capital outlay. With e-scooters being integrated into the transport system, this gives citizens greater accessibility to cheap transport.

E-scooters can also improve transport access in areas where public transport is limited. E-scooters provide the opportunity to bridge the transport network, allowing residents in areas where there is little or no public transport to travel into the city centre for a reasonable fee.

Community benefits

In the UK, 58% of car journeys are fewer than 5 miles (8 km), and in urban environments 69% of car journeys are fewer than 3 miles (5 km). Similar trends prevail in other European cities, and given the typical distances of e-scooter trips of 1-5 miles (1.5-8 km), there is a clear opportunity for e-scooters to replace car journeys within cities and help ease congestion.

E-scooters are ideal for congested city centres due to their compact design and door-to-door operability. Manoeuvrability through traffic, with typical legal speeds throughout Europe of 20 kph (12.5 mph), means that short journeys can be much quicker than private car travel in city centres, with the additional ability of being able to park e-scooters at the destination due to dockless systems.

In Paris, the average inner-city speed for the last mile of a journey was found to be 16 kph (10 mph), clearly demonstrating the possibility of e-scooters providing a faster form of transport than private cars.

As e-scooters become more widely adopted, congestion should begin to ease in city centres, decreasing journey times for private cars. However, as e-scooters and other forms of micromobility increase in popularity, city planners should reallocate road space from cars to micromobility, ultimately maintaining the status quo of slow inner-city speeds for private cars.

By providing a shared service, rather than the individuals purchasing a vehicle privately, city authorities can have greater control over vehicles, including maintenance of vehicles, applying speed limits, and creating geo-fenced areas, restricting where e-scooters can be used. A shared service allows those who cannot afford their own vehicle to access this form of transport for relatively little cost, as well as creating a more joined up and integrated transport network throughout the city.

Environmental benefits

E-scooters now have typical lifespans of up to two years in service, compared to 6 months when they first appeared on the streets of the US, thanks to improvements in design and battery technology, according to operators. An increasing number of operators use replaceable batteries, which limits the overall distance of workers who recharge the vehicles, and ensures that vehicles stay in the field, increasing their usage and reducing their greenhouse gas emissions per kilometer ridden.

Additionally, with the introduction of replaceable batteries, large diesel vans that were once required to transport the vehicles to and from recharging sites are no longer required, with operators starting to implement sustainable and low emission cargo bikes. The carbon footprint of e-scooters has rapidly reduced since their initial implementation, as demonstrated in Figure 9 produced by EY on behalf of Voi Technology, showing a 70% reduction in carbon dioxide (CO_2) emitted per km, down to 35 g CO_2 per km since January 2019.

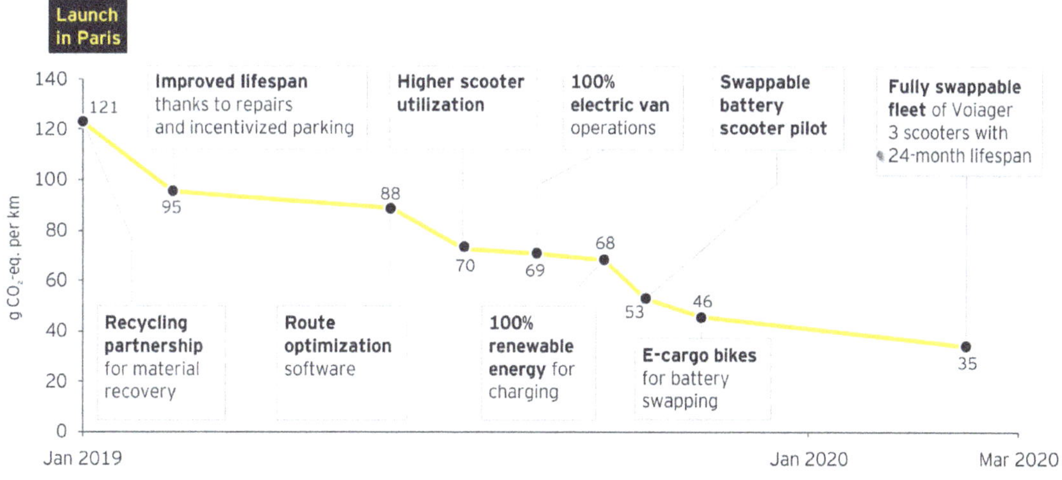

Figure 9 The reducing carbon footprint of e-scooters

Improving the environmental performance of the vehicles can help maximise carbon emissions benefits, and even small innovations, such as route optimisation and making the vehicles more recyclable, can be effective.

In addition to carbon savings, when e-scooters are used instead of privately owned cars, this reduces the overall air pollution caused by transport. Polluting gases such as nitrogen oxides and particulate matter are not produced by electric vehicles. Cleaner air in cities can have a large, positive impact on people's health, reduce the likelihood of lung and heart conditions, and save costs for health services.

Can you think of other benefits of e-scooters?

In the next section you will look at the limitations of e-scooters.

2.2.2 Limitations

Life expectancy

When operators first deployed off-the-shelf commercial products, they were not originally fit for purpose – neither for sharing, nor riding on rough cobblestone pavements as part of their daily routine. As a result, the first generation of e-scooters had a life expectancy of just a few months ⬛ not enough time for their operational GHG savings to outweigh their embedded emissions.

Since then, life expectancy has improved. However, in order for operators to have an economically sound business case, the majority of operations need to be conducted in urban, densely populated areas. There are very few operators running in rural areas, where there is a lower density of people, thus excluding those who live in these regions where the transport network is already limited.

Associated emissions

When e-scooters replace active travel (walking and cycling) over car use, this will increase overall CO_2 emissions rather than subtract from the overall transport carbon footprint. If appropriate measures are not used to discourage this form of modal shift, then e-scooters are in danger of becoming a less sustainable option than the current status quo.

Battery recharging is another important issue to consider. E-scooters are typically collected by larger vehicles for their batteries to be recharged, usually during off-peak hours, at night. They are spatially dispersed due to the nature of their usage, which generates irregular and long travel patterns for those collecting the vehicles. During the initial deployment of e-scooters by operators, little thought was given to controlling the method of this collection, with diesel vans often being the preferred method for gig-workers who charge the vehicles. The methods of collection are often non-rationalised, meaning collection routes do not follow an optimised route and so extra energy and fuel is used unnecessarily.

Safety

Many operators are aware that cities are demanding sustainability and safety from mobility solutions, and they have acted: improving the lifecycle emission performance of their products through more robust design of components and improving geo-fencing to nudge users towards parking responsibly.

The safety of users has been previously discussed in section 2.1, but it is worth highlighting that e-scooters represent a potentially dangerous form of transport for users sharing road space with large, motorised vehicles. In addition to safety concerns caused by other road users, the vehicles themselves also present a safety concern due to their inherent design. Typically, e-scooters have smaller, less robust wheels than classic forms of mobility (such as bikes), meaning that cracks and potholes represent serious hazards to users as they ride over them, causing instability. These are much more prevalent on the edge of roads and in existing cycle lanes where e-scooters operate.

Reliability

Unreported damage can also be an issue with e-scooters, meaning that a user could approach a vehicle and unlock it before finding out that the vehicle is not functional or safe to ride. This can cause frustration for users and create a negative experience of the operation and discourage users from engaging with this mode again. Even when damage is reported, the vehicles are often present on the streets till the end of the day when they're collected. This, again, can cause frustration for users who can see a vehicle that they are not able to use.

Accessibility

One final limitation, which is obvious to anyone not yet using them, is barriers to access through technology and physical fitness. At present, if you don't have a phone capable of downloading the app the operator uses or have internet access, then you aren't able to use these vehicles. This automatically excludes both poorer sectors and the non-technically-friendly, whom e-scooters could otherwise help because they are cheap to use.

Similarly, there is a lack of appeal of these vehicles for all of the older population and some people with disabilities. This is an issue with other forms of micromobility, as well as e-scooters: they can only be used by the physically fit, and nobody with walking ailments, neuromuscular diseases and visual impairments. These people will continue to prefer the safety of a more substantial vehicle.

From what you have learned, for which sectors of the population do the benefits of e-scooters outweigh the limitations?

2.2.3 E-scooter roadmap

It is important to understand the future landscape of the transport network in order to inform the decisions made today. Although forecasting technology developments is not an exact science, there are clear indicators today that help inform the likely trends of future shared mobility transport.

Through understanding the timeline for technology advances and changes in operation, cities can ensure that any policies and strategies devised are sustainable for the long term.

The roadmap in Figure 10 displays an indication of the likely route that e-scooters may take through to 2030, with the text below explaining the likely reasons for these changes.

Technology

Improving the lifespan of e-scooters is the main priority for operators currently and continuing for the next few years. This will involve new designs and improvements of critical components — such as wheels, frames and suspension systems — as well as advances in material technology. Self-reporting damage sensors, already present on some models, will become commonplace and aid in extending the life of e-scooters.

Figure 10 Roadmap of possible future developments of e-scooters through to 2030

As lifecycles improve, it is likely that e-scooter operators will focus more heavily on safety aspects, both for riders and pedestrians. Likely developments will include pavement and pedestrian detection systems, as well as technologies to prevent users riding whilst intoxicated.

Though it is hard to project what technology advancements for e-scooters will occur after 2025, this period aligns with likely advances and commercialisation of solid-state batteries. The introduction of solid-state batteries to e-scooters could help increase range as well as battery life.

Operations

Operators typically commission third-party gig-workers to collect e-scooters that are low on charge in the evening, and transport them to a facility where they can be recharged overnight. For the most part, operators do not control the method of collection, with the favoured collection method being diesel vans. This collection method is likely to remain for the next few years, though some operators are already making radical changes to their systems.

With the introduction of swappable battery technologies, this removes the need for gig-workers to transport scooters in vans; instead they can use other forms of low carbon transport, such as cargo bikes, to travel between each e-scooter. Collection methods such as this, and the use of electric vans, will be one of the main operational changes between 2022 and 2025.

From 2025 onwards, standardised MaaS systems will spread throughout UK and EU cities, and e-scooters will be a part of this; one of the primary advances being a change in the management of data and its exploitation to improve the transport systems of cities.

Environment

Improvements primarily due to increased life of e-scooters will results in CO_2 savings per mile, with operational CO_2 savings dominating environmental benefits towards the middle of the decade.

Further development of end-of-life processes for vehicles may result in up to 99% of the vehicle being recycled and/or reused by 2030, reducing embedded emissions of the vehicles.

Coverage

E-scooter ride sharing is available in over 90 European cities as of March 2020, covered by around 20 operators. These are primarily large cities, and will continue to be the primary deployment sites for operators as they seek profitable operations across the UK and EU.

As operators become profitable, e-scooter deployment will spread to smaller cities with populations of around 100,000 people, as operators continue to expand towards 2025. After the exploitation of smaller cities, e-scooter providers may look to solve the first and last mile journey challenges outside city centres, targeting the peri-urban areas of cities where a large proportion of citizens live.

Economics

At present, few e-scooter operators are profitable, primarily due to the short lifespan of vehicles: they need to be replaced every 6-12 months. Advances in the life of critical components will assist the potential profit-ability of e-scooter operators, and it is likely that the majority of operators will have a profitable business case by 2023.

As operations become profitable, this will propel the spread of vehicles to previously challenging markets, such as small cities and peri-urban areas. This will prove a critical time for operators, and could squeeze the number who survive: larger operators can offer less expensive journeys in challenging environments, while smaller operators will either fold or be bought out.

How does the roadmap align with what you anticipated in the previous section?

Was anything surprising to you about the likely development of e-scooters?

2.3 E-bikes

2.3.1 How do they work?

Shared e-bikes are a form of micromobility that has developed off the back of hugely successful bike sharing programmes. E-bike services typically use the same model as standard bike sharing programmes, and came to fruition around 2016, with both docked and dockless bikes now available throughout various European cities.

E-bikes have the potential to attract a wider range of users than non-motorised bicycles, whilst still providing health benefits and a low-cost, environmentally-friendly form of travel. So far cost has made them less socially inclusive than conventional cycles, but more inclusive by age and fitness.

Figure 11 An electric bike charging in a docked station

E-bikes typically rely on an app for use. Most of these apps require a monthly or annual membership fee, plus separate usage fees. Users unlock the bike through an app at a cost of around 1-2 £/€, depending on the operating model. ncl either gives them use for a set amount of time, typically 30 minutes to an hour, or a charge may be incurred on a per minute use case. If the bike is in use longer than this initial period, further charges may apply based on the time in use. Some schemes may include rental options for the day, allowing users to explore areas further afield, as well as weekly or monthly subscriptions.

Due to their ability to assist users in their efforts to 'turn the pedals', e-bikes are able to attract a much wider range of users than non-motorised bicycles, whilst still providing health benefits as well as a low-cost, environmentally-friendly form of travel. Their cost makes them less socially inclusive than conventional cycles, but they are accessible to a wider range of ages and physical fitness.

E-bikes can also help users to access and experience areas Beyond the city limits, ncluding mountain trails or rural exploration with particularly challenging terrain.

Figure 12 Electric mountain bikes for exploring hilly regions

E-bikes hold a clear advantage over regular bike sharing schemes in rural areas, where typical cycling distances are longer; in regions with difficult terrain, such as hills; and to increase the mobility of older and less fit people who are no longer able to ride a traditional bike.

While e-bikes are still in their relative infancy, the business model for bike sharing is well known and understood and could lead to easy expansion of e-bike schemes over the coming years.

The common operating model for providers is to place e-bikes in central locations close to city amenities, other transportation hubs (such as train stations), or in high footfall areas. Throughout the day their electrical charge decreases from use, and dependent on the operating model used, they can be either recharged in the docking bay, provided by an electrical connection, or recharged over night by gig-workers in the same way as e-scooters.Unlike scooters, e-bikes tend to make use of docking stations, both for security reasons and due to their larger size and instability when parked. This also makes provision of electrical charging facilities simpler.

Like many forms of transport, e-bikes have positive and negative effects on cities. It is crucial that authorities understand these in order to limit the negative impacts and ensure they work for citizens, first and foremost, and the operators.

Case study: E-bikes in Berlin

Jump, formerly part of Uber and now Lime, launched their first shared e-bikes in Berlin in 2018, adding to the non-motorised shared bikes already present in the city. An initial 1,000 vehicles were deployed throughout 12 districts in the German capital.

The e-bikes can be rented via the Uber app, which can be downloaded free of charge. Available bikes can be found via an integrated GPS sensor and the app. If the desired electric bike is reserved via the app, the customer is sent a message with a code to unlock the bicycle. The PIN is entered via a keyboard on the rear wheel, which unlocks the dock and you are ready to go.

The Jump bikes cost 1 euro for the first 20 minutes, then 10 cents for every additional minute. The time starts running as soon as the e-bike has been reserved via the app. As part of the hire scheme, the following policy is also applied allowing maximum flexibility during trips:

→ The bikes can be parked for an hour in between.

→ There are no costs during this parking time.

→ If the bike is parked outside the service area, there will be an extra charge of 25 euros.

Once the user has finished with the electric bicycle, they may then park it at any bicycle stand within the Jump service area and the bike is secured with a shackle lock. There is no requirement to park the bicycle in the same location it was removed from.

What do you see as the potential benefit of e-bikes over conventional shared bike schemes? Would they be useful in your area, and if so, why?

KEY DEFINITION:

E-bikes – motorised bicycles with an integrated electric motor used to assist propulsion. E-bikes fall into two categories: 1) Pedal-assist / pedelec-bikes that only add motor assistance when the rider pedals; 2) Throttle bikes that allow the bike to be powered entirely from the motor. Both retain the ability to be pedalled by the rider and are therefore categorised separately to electric motorcycles. In this book the term e-bike will be used to refer to the first category only.

2.3.1 Applications and benefits

Applications

Shared e-bikes are a first and last mile mobility service which can be used to link up other transport services and aid in creating a fully integrated transport network. In view of their familiarity, their simplicity and comfort, e-bikes are more likely to replace longer journeys than other forms of micro-mobility, such as e-scooters. This potentially makes it much more likely that e-bikes will replace full car journeys, rather than walking or public transport.

Economic and social benefits

E-bikes, like e-scooters, can benefit citizens of any socio-economic status. Unlike passenger cars, where large amounts of capital investment are needed or expensive lease schemes, shared e-bikes can cost as little as £2/€2 for 30-60 minutes of use with no capital outlay. Through the integration of e-bike services into the transport system, this gives citizens greater accessibility to cheap transport.

E-bikes are highly accessible vehicles: if you know how to ride a normal bike, then no training is required to use them and no driving licence is required either. They provide added accessibility over regular shared bikes due to their pedal-assist batteries. This allows greater access to those who may struggle to ride long distances or over difficult terrain, attracting new people to cycling. Those who may already cycle will have the benefit of being able to travel further with less effort than a regular bike, increasing the likelihood of them selecting an e-bike for a moderate journey over a private car.

Community benefits

As with e-scooters, e-bikes are ideal for congested city centres due to their compact size and ability to use cycle lanes. Manoeuvrability through traffic, with typical legal speeds throughout Europe of 25 kph (16 mph), means that short journeys are much quicker than private car travel in city centres.

A shared service allows those who cannot afford their own vehicle to access this form of transport for relatively little cost, as well as enabling them to follow an integrated transport route through the city. Also, the operators control the vehicles, including their maintenance, specifying speed limits, and creating geo-fenced areas where they can be used.

Figure 13 A row of e-bikes parked up ready for use

When e-bikes are utilised as part of a docked system, this results in vehicles being parked in designated areas where they will not be a trip hazard to the general public or a potential eyesore. A docked system also allows for a potentially integrated charging system. This removes the need for staff to either collect vehicles for recharging or swap batteries on vehicles overnight.

Environmental benefits

In section 1.1.1 you learned about the benefits of shared mobility and compared the environmental impact of different transport options in Figure 1. It was seen that e-bikes produce the next lowest emissions, after the metro, for assisted powered travel (walking and cycling are classified as active travel).

Provided e-bikes do not replace active travel journeys, these vehicles have the largest potential to reduce carbon dioxide than any other form of shared mobility.

Health benefits

Unlike the two previous forms of shared mobility you have looked at, e-bikes offer a health benefit as well as an environmental one (though not for those upgrading their regular bike for one). E-bikes work on a pedal assist basis, meaning that as you pedal and put effort in, the battery can boost the power you make. The important thing here is that the rider is still doing the majority of the work, and so can experience the health benefits of exerting energy and exercising. Research from Bristol University showed that using an e-bike burned more calories per mile travelled than walking, but less than a conventional pedal cycle.

Can you think of any other benefits of e-bikes?

After that, we will look at the limitations of e-bikes.

2.3.3 Limitations

Docked v Dockless

Unlike e-scooters, there is no preferred way of operating these vehicles, whether it be a docked or dockless system, and whether a docked system also charges the vehicle.

Docked systems provide limitations on the ability for e-bikes to be a first and last mile form of mobility, as you are only able to travel to where a dock is, and not your final destination. Though there are clear benefits to a charging docking station, as discussed in the section about community benefits, there is a cost to this infrastructure as well as suitable space required for installation (something that is a premium in city centres). A docked system may also limit the ability for these vehicles to provide a true last mile travel service, potentially acting as a barrier to use.

Dockless systems on the other hand can provide true door-to-door travel allowing full journey completion on a single mode of transport. Though a dockless system removes the need for infrastructure, the vehicles can become spread out throughout the city. E-bikes are significantly harder to transport to and from charging warehouses than e-scooters, due to their size and weight, and so battery swapping technology will have significant environmental and economic benefits, though the logistics remain more difficult than a docked system.

Replacing active travel

One of the main threats to this model is that e-bikes are more likely to replace other forms of active travel (walking and regular bikes) rather than replacing car use; and this shift would therefore increase overall CO_2 emissions, rather than subtract from the overall transport carbon footprint. It would be difficult to discourage this direction of modal shift, so in the short term e-bikes might become a less sustainable option than the current status quo.

While an inability to ride a bike is an obvious limitation for the use of e-bikes, less obvious is the fitness level required for even these supported vehicles. The majority of e-bikes are pedal assist, so the user is required to exert energy and effort alongside the electric motor; this will exclude some older and less fit users, especially if considering longer journeys or undulating terrain.

Reliability

Unlike e-scooters, which have solid wheels, e-bikes have inflated tyres and are susceptible to flat tyres and punctures rendering them unusable. Similarly, having brakes that function well and correctly can be a key safety concern for riders. So, while a docked system can remove the need for staff to collect and recharge vehicles, they are still vital for maintenance and upkeep of vehicles, which is critical for their use. If these are unreported, a user could approach a vehicle and unlock it before finding out that the vehicle is not functional or safe to ride. This can cause frustration for users and create a negative experience of the operation and discourage users from engaging with this mode again.

Other safety issues include the use of e-bikes on the roads with other road users, including large vehicles such as buses and lorries. It is irrelevant that bicycles have been around longer than motorised transport; there are still issues surrounding visibility, speed, bulk and the appropriate behaviour of motor vehicle drivers towards any kind of cyclists. Bike lanes and reducing road speeds may support the safety of e-bike users, but they are not universal and are often disregarded.

Unregulated deployment

Finally, without clear guidance on the distribution and provision of e-bikes by cities through regulation, we may get too many of them. This can lead to inevitable destruction, such as in Xiamen (Fujian province, China) in 2018. No regulation was in place for the undocked systems, causing bikes to be over-provided leading to abandoned vehicles and crowded streets.

From what you have learned, do you believe that the benefits of e-bike schemes outweigh the limitations?

2.3.4 E-bike roadmap

It is important to understand the future landscape of the transport network in order to inform the decisions made today. Although forecasting technology developments cannot be an exact science, there are clear indicators today that help inform the likely trends of future shared mobility transport.

Through understanding the timeline for technology advances and changes in operation, cities can ensure that any policies and strategies devised are sustainable for the long term.

The roadmap in Figure 14 displays an indication of the likely route that e-bikes may take through to 2030, with the text below explaining the likely reason for these changes.

Figure 14 Roadmap showing the possible future developments of e-bikes through to 2030

Technology

Robust, low-maintenance bikes have been prevalent for a long time, with many years of development already having taken place for non-motorised bike sharing schemes. As such it is unlikely that technology advancements in this area will be the priority for operators.

Cargo e-bikes may see the greatest development over the coming years, with the industry still relatively new and uptake of vehicles likely to increase rapidly as low emission zones and clean air zones are introduced in cities, making typical delivery methods (diesel vans) challenging for businesses with limited capital to purchase new, compliant vehicles.

Towards the end of the decade, e-bike providers may look at larger battery capacities, either through advances in energy density of li-ion batteries, or moving over to solid-state batteries. This could allow users to hire e-bikes for a full weekend without the need to recharge.

Operations

Pay-as-you-ride services are likely to dominate for the coming years, as a system that has proved popular with current non-motorised shared bikes. Similarly, a mixture of both docked and dockless systems is likely to be available, and can be decided on a case-by-case basis.

Moving towards 2025, subscription services are likely to become more prevalent and be popular with urban environments for commuting short distances.

From 2025 onwards, MaaS systems will be much more prevalent throughout UK and EU cities, and e-bikes will be a part of this; one of the primary advances being a change in the management of data and its exploitation to improve the transport systems of cities.

Environment

As an industry that is highly developed, it is unlikely that any environmental improvements will come from the bike itself. Battery production and management of batteries to extend their lives are likely to form the main environmental improvements in the early part of the decade.

Alongside battery improvements in CO_2 footprints, operators will look to control the end-of-life of their vehicles more closely, especially after recent media backlash due to aforementioned mass scrapping of bikes in China. As well as reducing embedded emissions, by recycling and reusing parts of the e-bikes more sustainably, this will also help contribute to the profitability of operators. It is likely that by the end of the decade the lifespan of e-bikes will be greater than 5 years.

Coverage

Major cities and densely populated areas will continue to be the focus for e-bike schemes, following a similar coverage and deployment method to already existing non-motorised bike schemes. This will continue through to the middle of the decade, when e-bike operators will probably look to deploy additional schemes in smaller cities within riding distance of popular rural attractions. A primary focus of this will be targeted journeys such as mountain trails, aligning with further technology developments of this style of bike.

Economics

E-bike operators will continue to be profitable in large cities and tourist destinations where similar non-motorised bike schemes are already deployed, replacing some of these vehicles as well as adding to the network. Both pay-as-you-ride and subscription services have been shown to be profitable, though some schemes do require public funding and/or sponsored services that act as an advertising platform.

By the middle of the decade the majority of deployment sites will be profitable from rider revenue income alone, removing the need for any public funding of these schemes. This will allow an expansion into previously challenging markets, such as small cities and rural regions.

The future of non-electric bike sharing

This form of shared mobility is fully mature, and so no major advances are expected for non-electric bike sharing. If MaaS apps mature and become more widespread, it is likely that integration into these systems will be the main development for this transport service.

How does the roadmap align with what you considered in the previous section?

Was anything surprising to you about the likely development of e-bikes?

2.4 Part 2 Quiz: Test what you've learned

Well done – you have reached the end of Part 2!

So far, you have focused on what shared mobility is and the different vehicle options available. You have looked at their potential applications, benefits and limitations, as well as peering into the future.

It's important to remember what you learned in Part 2, so here are the top three takeaways to remind you what was covered:

KEY TAKEAWAYS

1. Micromobility refers to a range of small, lightweight vehicles operating at speeds typically below 25 km/h and driven by users personally. Micromobility devices can include bicycles, e-bikes, e-scooters, electric skateboards, electric hoverboards and Segways.

2. Micromobility should primarily enable first and last mile journeys to connect with other forms of sustainable travel, or replace entire journeys that would ordinarily be covered in a private car. Cities and operators should work together to identify deployment sites that promote shared mobility travel over more carbon-intensive modes, rather than compete against other sustainable transport services.

3. If cities aren't prepared with appropriate policy and regulation, the deployment of micromobility by operators can lead to issues such as over-deployment, dangerous use, safety concerns and street clutter. Local authorities therefore need to take an active role in the deployment of micro-mobility through tenders, parking regulations, speed limits, insurance and environmental sustainability.

In Part 3 we will be exploring the applicability of these shared mobility options in the real world, and what the key considerations are for successful deployment. We will also discuss the concept of mobility hubs and how data can be valuable not just to operators but to cities as well.

But before you carry on, try this short, multiple-choice quiz:

1. Which of the following is NOT a form of micromobility?

1. Electric scooter

2. Electric hoverboard

3. Electric milk float

4. Pedal bike

2. What is the typical range of an e-scooter based on a single charge?

1. 5 miles (8 km)

2. 10 miles (16 km)

3. 40 miles (64 km)

4. 200 miles (320 km)

3. Which of the following does not provide a safety hazard for e-scooter users?

1. Bike lanes

2. Pot holes

3. Drain covers

4. Fast-moving traffic

4. Which of the following is not considered a benefit for a docked e-bike system?

1. Vehicles are in fixed locations allowing easier and faster maintenance.

2. Vehicles are parked in designated areas reducing trip hazards.

3. Vehicles have the ability to recharge if a grid connection is present.

4. Vehicles are in fixed locations aiding door-to-door travel.

5. In which of the following areas are e-bikes unlikely to be deployed?

1. City centres

2. Mountain trails

3. Shopping malls

4. Popular rural attractions

You can find the answers to this quiz at the end of the book.

HOW IS THIS APPLIED TO CITY PLANNING AND MOBILITY HUBS?

In Part 3 you will now explore how this knowledge can be applied in the real world. This includes where we should deploy these vehicles for maximum environmental benefit and uptake, as well as what cities can do to support this. Then finally we will look at the concept of mobility hubs and how data is changing the way of the world and helping the shared mobility industry grow.

3.1 City planning

"Great design is a multi-layered relationship between human life and its environment."
– Naoto Fukasawa

City planning for transport is key to creating a network that is fit for purpose and serves its citizens. This includes how the roads are utilised for transport as well as the wider landscaping for pedestrians and how they interact with transport.

Further to this, city planning should incorporate environmental policies and targets that will have an influence on decisions made regarding transport. Increasingly, cities are being encouraged to create master plans and sustainable urban mobility plans (SUMPs) with the environment and carbon reduction at their heart. As more countries and cities target net zero emissions in the future, these plans are key to making sustainable choices now in order to realise their future targets.

Sustainable Urban Mobility Plan (SUMP)

A SUMP is a planning concept created and applied by cities and local authorities for strategic mobility planning. Its principal aim is to improve accessibility and provide high-quality, sustainable mobility for the entire urban area. This includes supporting the integration and balanced development of all modes considered sustainable. A SUMP is key to solving urban transport problems, helping to steer the direction of city planners when making choices about new forms of mobility.

SUMPs consider the whole urban area with cooperation across different policy areas, transport modes and local residents. It ensures that a variety of sustainable transport options are available for residents. The aim is to create the safe and healthy passage of people and goods across the city, with consideration for fellow residents and the urban environment.

A summary of the key principles behind SUMPs is stated below:

➜ **Plan for sustainable mobility:** Sustainable mobility should reduce air and noise pollution, greenhouse gas emissions and energy consumption. It should also enhance the attractiveness of the urban environment, quality of life and public health, whilst improving road safety and security. The sustainable mobility network should be accessible and meet the basic mobility needs of all users, regardless of socio-economic status. While meeting the sustainability needs of the city, the plan should also balance this with economic viability, optimising efficiency and cost-effectiveness.

➜ **Cooperate across institutional boundaries:** A high level of cooperation, coordination and consultation is required across different levels of government and between institutions (and their departments) in the planning area. Additionally, coordination with public and private sector providers of transport services is required to ensure a joined-up approach.

➜ **Involve citizens and stakeholders:** The plan should focus on meeting the mobility needs of people in the area, both residents and visitors, as well as institutions and companies based there.

➜ **Assess current and future performance:** It should provide a comprehensive review of the existing situation and establish a baseline against which progress can be measured.

➜ **Define a long-term vision and a clear implementation plan:** The plan should be based on a long-term vision for transport and mobility development for the entire area and cover all modes of transport: public and private; passenger and freight; motorised and non-motorised. It should also include infrastructure and services with an implementation plan that has a timescale, budget and clear allocation of responsibility.

→ **Develop all transport modes in an integrated manner:** Though each part of a SUMP will be looked at individually, a wider vision is required for how all the services will work together. The plan should put forward an integrated set of measures to improve quality, security, safety, accessibility and cost-effectiveness of the overall mobility system.

→ **Arrange for monitoring and evaluation:** The implementation of a SUMP must be monitored closely. Progress towards the objectives of the plan and meeting the targets should be assessed regularly, based on the chosen performance indicators.

→ **Assure quality of service:** A SUMP is a key document for the development of an urban area. Having mechanisms in place to ensure a SUMP's general professional quality and to validate its compliance is key to ensuring that the best possible vision of the region is taken forwards and implemented.

What do you think are the key benefits of a SUMP, and why would it be useful for your own city to develop one?

3.1.1 What cities need to take into consideration

Good practice in the deployment of shared mobility is crucial to maximise the benefits of their deployment. The following key considerations should be looked at when deploying these vehicles:

→ **Cooperation:** Local authorities and private operators have to establish how to integrate shared mobility services into the city's transport system. Public transport and active travel should be the backbone of sustainable urban mobility, with shared micro-mobility services supporting these modes rather than competing against them. This should be communicated to operators and dictate where shared mobility services are placed to ensure they contribute to the city's sustainability objectives, and not just for the operators' business case.

→ **Public spaces:** Cities and towns have historically allocated a large proportion of public space to privately owned vehicles. This bias towards private cars reduces the potential for deployment of more sustainable modes. The reallocation of space for the safe use of active travel and shared mobility is not about penalising private vehicle use; it is about creating a network that allows more sustainable modes of travel to be a genuine option for those who want to use them, whilst feeling safe and seen by other road users.

→ **Infrastructure:** Appropriate infrastructure is required in order to deploy successful shared mobility, whether this be bike racks, charging bays for EV car club vehicles, or designated parking for e-scooters. Converting car parking spaces for shared mobility provision should be considered, rather than taking away further pedestrian space; and locations for deployment should provide good visibility of the service: a hidden-away car park with little footfall is not a good location. Reallocating parking spaces is relatively fast and cheap, with only signage, road paint and barriers needed, along with the possible addition of a chargepoint for car club vehicles, if required. Additionally, the decreased parking allocation for cars may increase demand for shared mobility. Dockless e-scooters can still raise challenges for parking provision at the end of a journey that is not in a high footfall area, as infrastructure cannot be provided at every location to which users want to travel. Mixed solutions must be considered so that dockless systems do not result in excessive cluttering of public space. This could involve requiring e-scooters to be parked in designated areas in dense, highly populated areas, while full free-floating is allowed in others. This should be discussed in depth with operators, to ensure that GPS systems on the vehicles have enough precision to enforce these regulations.

→ **Fees:** In general, local authorities charge private services to use public space to conduct their business. Fees charged to e-scooter operators could feed into the general city budget, or be spent on relevant measures such as investment in infrastructure and traffic calming measures.

→ **Safety:** Micromobility customers can be considered vulnerable road users, with vehicles often sharing the same road space as private cars, vans and HGVs. User-centred safety solutions such as helmets, high visibility vests and reflective bands can help protect users. However, the greatest threat to the safety of riders is other road users, and their danger can outweigh these solutions. Actions such as reducing speed of traffic, creating micromobility lanes and introducing traffic calming measures are generally more beneficial and more far-reaching.

→ **Social and economic demographics:** One of the obvious benefits of shared mobility is the elimination of upfront capital investment, allowing users to pay for their journey and not the vehicle itself. Therefore, deployment of these vehicles should take into account lower-income areas that are often deprived of cheap, low emission transport options. This levelling up can help bridge the gap between economic and social classes, providing affordable low emission transport with no capital cost.

→ **Location:** Each form of shared mobility should be assessed separately when considering the location of deployment. For shared micromobility, it is important to note that these modes of travel should primarily enable first and last mile journeys to connect with other forms of sustainable travel, or replace entire journeys that might otherwise be covered in a private car. Local authorities and operators should work together to identify deployment sites that promote shared mobility travel over more carbon-intensive modes, rather than compete against other sustainable transport services. We will explore locations for deployment further in the next step of this course.

3.1.2 Placement of shared mobility

The introduction of shared mobility into a city as part of a transport system should primarily encourage sustainable travel. For car clubs, this means conveniently placed vehicles to allow the general public to consider giving up their own personal vehicles. For micromobility, this means enabling first and last mile journeys to and from other forms of sustainable travel, or replace entire journeys that might otherwise be made in a private car.

It is therefore imperative that cities and operators work together in order to identify deployment sites that promote shared mobility travel over more carbon-intensive modes; shared mobility should not compete with transport services that are already low emission, such as public transport or active travel.

Each city is different, and so the responsibility of site identification should lie primarily with the city. Operators coming into any city will not have in-depth local knowledge of the nuances of the transport system or wider initiatives that can affect travel. Cities need to share knowledge with operators so that locations can be identified where the operator's business case will stack up and where the city can benefit most, filling gaps in the transport network and ensuring that a high proportion of journeys are replacing car journeys. There are, however, some sites that should be explored common to most cities.

Train stations

When you walk out of a train station, often the first thing you see is a long line of taxis waiting to take travellers on the next part of their journey. Placing micromobility at train stations is ideal for travellers with no luggage, as alternatives to the start and end points for taxis. Typical train journey distances range from 10 to 200+ miles, so there is little likelihood of micro-

Figure 15 E-scooters from different operators in a dedicated parking bay

mobility services replacing train journeys, but these two modes of travel do complement each other well.

Of course other transport services are also located close to train stations, such as bus stations, so the placement of shared mobility at train stations should be prominent to aid travellers' choice between public transport and shared mobility.

Bus stations

While bus services and micromobility both offer shared transport facilities, co-locating these services at bus stations will remind travellers that e-mobility can provide first and last mile travel specificity, whereas buses on their set routes cannot. Whether or not users of micromobility are replacing their active travel to and from bus stops and stations, the efficient coordination of bus and micromobility should encourage private car users to consider switching to a more sustainable way of travelling.

Most bus routes probably extend beyond the operational boundaries of micromobility, so this is another reason that co-locating these services will enhance their complementarity and even persuade non-users to reconsider.

Additionally, the value system of bus users probably overlaps with that of micromobility users: those who don't use cars for economic and environmental reasons. Therefore, co-location will help to promote micromobility to those most in favour of them.

Technology industrial parks and universities

E-scooters were first deployed in tech-savvy neighbourhoods and universities on the west coast of America; these remain areas that have the highest utilisation of e-scooters in the world. Due to the climate on the west coast, and the typical social demographic at universities and technology industrial parks, these citizens are still the ideal target market for e-scooters and other forms of shared mobility.

With limited parking on most campuses and an eye towards sustainability, university leaders have always encouraged their students to use alternative forms of transport. Banning cars and providing efficient and economical shared mobility should ensure that students gain the habit of alternative transport modes.

Figure 16 E-bike provision on a university campus

Retail hubs

Retail hubs are high footfall areas, usually the most frequented by citizens from throughout the city and its surrounds. They therefore present an opportunity to make shared mobility visible to many people who might otherwise not be aware of such services.

Given the frustration of traffic congestion in reaching city centres, placing micromobility prominently alongside retail hubs is the ideal way to change the mindset of citizens to consider these vehicles as the ideal way to make short journeys to reach the city centre.

Residential areas with no off-street parking

Densely populated urban environments, while attractive places to live, don't often allow for private parking. Those who do own vehicles will either have to pay for parking permits, or may struggle to park close to their homes.

Car clubs placed at these locations, with designated public bays, can reduce the need for privately owned cars and remove the stress and extra cost associated with owning vehicles in these areas. The high density of people in urban environments provides a viable business case for operators.

Residential areas with a low uptake of private vehicles

A low proportion of private vehicles is not always caused by limited parking. It could also be a sign of depravation in a specific area, with residents unable to afford private vehicles due to their economic status.

By providing car clubs and micromobility in these locations, cities can ensure that their residents are not limited in their travel opportunities by their income. Shared mobility vehicles do not require any capital investment and residents can simply pay for the journeys they make.

3.1.2 City planning roadmap

It is important to understand the future landscape of the transport network in order to inform the decisions made today. Although forecasting technology developments cannot be an exact science, there are clear indicators today that help inform the likely trends of future shared mobility transport.

Through understanding the timeline for technology advances and changes in operation, cities can ensure any policies and strategies devised are sustainable for the long term.

The roadmap in Figure 17 displays an indication of the likely route that cities may take through to 2030, with the text below explaining the likely reasons for these changes.

National objectives

As more countries sign up to net zero targets for emissions, responsibility will be placed on cities to provide more low emission transport options. This will be with an aim to reduce private vehicle use in city centres and transfer to active travel, public transport and electric shared mobility. As this begins to take hold, more ambitious targets, such as reducing the total number of privately owned vehicles, will come into action. Cities will look to build shared mobility as one of the cornerstones in their sustainable urban mobility plans, with a joined-up and supportive role for larger-scale low emission public transport networks.

Funding

Over the next few years, it is expected that governments and cities will begin to offer subsidies and incentives for shared mobility operators to deploy vehicles in their regions. As with any new technology, the business case for shared mobility is currently challenging, so until the economic case for these vehicles improves, public sector investment will be key. By the middle

Shared Mobility Technology and Policy Roadmap

SuSMo

City Plan

	Development and Enabling				Transition Phase		Scale-up and Realisation of Benefits				
	2020	2021	2022	2023	2024	2025	2026	2027	2028	2029	2030
National Objectives	Reduction in city centre private vehicle use through providing alternative transport					Stabilise and reduce the total number of privately owned vehicles and build reliance on shared mobility					
Funding	Subsidies required for new shared mobility modes			Subsidising shared mobility may be required in some areas, passive funding may help increase uptake i.e. parking infrastructure & chargepoints							
Supporting Innovation	Trials being run in cities to understand the impact on the transport network					Private sector led trials, taking over from public sector					
Law and Policy	Policy centres around reducing car use through making low emission alternative modes more appealing & widespread					Policy shift to focus on disincentives for private vehicle use in city centres					
Public Perception	Public see shared mobility as novelty not an alternative to private car use				Positive shift in public perception as shared mobility becomes more widespread						
Intelligent Transport	Early MaaS systems trialled in a few cities			MaaS deployed more widely with multi-city integration			Building of initial infrastructure for future deployment of CAV				
City Landscaping	Creation of city wide bike lanes, parking, and EV car club charging infrastructure				Major plans to reallocate city road infrastructure including possible no drive zones in the city centre						

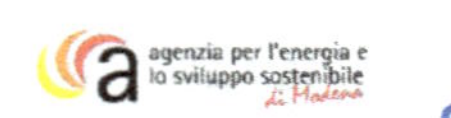

Figure 17 Roadmap of future city plans for shared mobility technology and policy

of the decade, it is expected that the majority of funding will come from the private sector, though some smaller cities and towns may still require public sector support. For larger cities, passive funding may still continue to help increased uptake. This may come in the form of providing parking infrastructure and chargepoint locations.

Supporting innovation

Due to the relative infancy of the industry, support for innovative solutions will be required from cities, with trials focusing on environmental studies and impacts on the transport network. As with funding, support for these trials will be taken over by the private sector towards the middle of the decade, as the economics of shared mobility is understood more and the technology becomes less disruptive.

Law and policy

Policy will concentrate on reducing car use through incentives for low emission alternative forms of transport, both public transport and shared mobility. Particular focus will be placed on making these modes more appealing for users and creating a full and complete network within the city. Once this is achieved, new policy may be implemented to focus on disincentivising private vehicle use in city centres. This may come in the form of reduced or more expensive parking, charging vehicles for entering city centres or low emission zones that hold a particular focus on disincentivising older, more polluting vehicles.

Public perception

At present shared mobility modes such as e-scooters may be seen as a novelty by the general public, rather than a plausible alternative to private car use. As increased education surrounding carbon emissions is realised and the general public are educated on the benefits of shared mobility, this perception may begin to change. As more cities deploy shared mobility, these modes of transport should begin to be a viable travel option to replace more polluting journeys such as private vehicle use.

Intelligent transport

At present there are a few early city demonstrators of Mobility as a Service (MaaS), enabling a single point of access to multiple modes of transport. If the MaaS industry starts to get a foothold in the market, then it is expected that this service will become more integrated. This may include neighbouring cities utilising the same platform, with the appropriate operators for each city buying into this service. Towards the end of the decade, connected and autonomous vehicles (CAV) should be more advanced, and trials will begin in a few leading cities. At this point cities will begin to take action and prepare for this technology, with the building of smart infrastructure that can aid in the deployment of CAV in the 2030s.

City landscaping

Passive funding of shared mobility will result in cities reallocating road space to these low emission transport options through the creation of more cycle lanes and increased parking for these vehicles. As electric vehicles become more prevalent, investment in chargepoints and electric vehicle parking bays will also increase, including specific parking bays for electric car club vehicles. Towards the middle and end of the decade, more ambitious plans may be implemented to reduce the number of privately owned vehicles in city centres. This could include zero emission zones, that prohibit any non-electric vehicles from entering these areas; or no drive zones, that reallocate city space to pedestrians and shared mobility users.

3.2 Mobility hubs

KEY DEFINITION:

Mobility hub – a recognisable place with an offer of different and connected transport modes, supplemented with enhanced facilities and information features to both attract and benefit the traveller. It should be designed and spatially organised in an optimal way so as to facilitate access to transport between modes. This can include human powered and electric shared modes, as well as providing extra transport-related and digital services, such as public transport and Wi-Fi connectivity.

3.2.1 Benefits and key considerations for mobility hubs

Now that you have been introduced to the concept of mobility hubs, it's time to understand the benefits and key considerations for mobility hubs.

Size of mobility hub

The size of a mobility hub should suit both the size of the potential location as well as provide the necessary services required by the community likely to use them. Three different sizes of mobility hub are given as examples below.

Figure 18: A small mobility hub in Paveletskaya Square in Moscow

→ **Small hubs:** These are generally deployed in densely populated areas and combine public transport links with shared micromobility (bikes and scooters). They can also act as potential links to medium-sized hubs.

➔ **Medium hubs:** This size of hub can connect suburban areas to inner city areas. They typically combine public transport links, car clubs, chargepoints and shared mobility. There is scope here to provide additional amenities such as cafes, toilets and delivery lockers.

➔ **Large hubs:** This size of hub can connect local, regional and national public transport routes, provide electric vehicle charging for long- and short-stay visitors, and shared micromobility. Additional amenities would be expected at these sites, such as cafes, toilets and delivery lockers, as well as the potential for further retail.

Benefits

➔ **Smarter sustainable transport planning**: Mobility hubs reclaim urban space for sustainable and equitable modes, reducing the dominance of the private car and associated problems of congestion, carbon emissions, air quality and social exclusion. A network of such hubs creates an attractive, integrated, viable alternative mobility lifestyle.

➔ **Convenience:** Mobility hubs provide convenience for multi-modal trips, with the possibility of seamless switches and improved links between different layers of transport, such as the core public transport network and shared services.

➔ **Choice of modes:** Mobility hubs provide choice for different journeys and needs. They encourage people to think multi-modally and are a complement and enabler of full digital integration of services – Mobility as Service. This, in turn, allows people to reduce their car use and the associated impacts.

➔ **Plugging the gaps in the public transport network:** In suburban areas they can perform a sustainable 'first or last mile' connection to the nearest bus or railway services, in a cost-effective way. They can provide flexible 24-hour services as a sustainable, accessible alternative to private car ownership.

➔ **Improving safety and comfort:** Mobility hubs, by design, offer a safer and more comfortable dwell time, leading to improved access for more vulnerable users.

→ **Improved accessibility:** It is possible for mobility hubs to provide space for adapted and inclusive modes as part of overall transport solutions.

Key considerations for the design and location of mobility hubs

When designing a mobility hub, there are a number of key considerations that need to be accounted for in order to successfully deploy mobility hubs. These are summarised below:

→ **Convenience:** How convenient the location of the mobility hub is from a user perspective. This includes the potential access to the location. Things to consider include: public transport routes, cycle routes, sites with limited parking and major transport interchanges (e.g. train stations, park & ride).

→ **Installation:** How feasible installation at the location will be. For sites that include chargepoints, a strong electrical supply is needed.

→ **Visibility and accessibility:** It is important that mobility hubs are clearly visible and accessible to maximise their use. Clear signage and wayfaring signs should be deployed at sites, with disabled access also important.

→ **Ease of switching modes:** It is important that users can switch to a MaaS system, so that they don't have to navigate a number of different apps when purchasing.

→ **Safety:** Quality paving, good street lighting, CCTV and safe road crossings will all ensure a high level of safety for user groups.

→ **Practical facilities:** At a minimum, a mobility hub should provide ticketing and journey information to users. Provision of other facilities should be considered to increase the practicality and visual appeal of mobility hubs, including seating, weather proofing, mobile network connection (or Wi-Fi), litter bins/recycling, toilets, community spaces such as cafes.

→ **Community consultation:** Co-design is important, particularly for hubs that aim to link in with community activities and become a community hub in their own right. Buy- in and sense of ownership is key.

Figure 19 An example layout of a large mobility hub

Planning guidance and considerations

When developing a mobility hub network, there are a number of key stake-holders that need to be involved to ensure the successful deployment of these services.

CoMoUK, a UK based non-governmental organisation (NGO) dedicated to promoting shared mobility, has issued guidance on the timeline and likely stage gates that should be followed when setting up a mobility hub network; this is shown in Figure 20.

Total time/ overlaps	Months	Activity	Notes
Months 0-2	2	Phase 1 consultation	Consultation to gain local support and identify location, carry out site visits and agree the key elements and design
Months 2-4	2	Design phase	Specification converted to technical design documents
Months 3-6	3	Phase 2 stakeholder engagement	Public promotion campaign and agreements in place with all stakeholders
Months 7-8	2	Tender service providers	This phase will depend on whether there are existing service provider contracts or new tenders are required. A batch of hubs may be done at once. This phase can only begin once specifics are known, e.g. EV charging, hub size, type of signage.
Month 9	1	Tender for construction	
Months 10-12 or 10-15	2-5	Building and public information	Time depends on availability of construction team, local services and local conditions.
Ongoing		Monitoring, report, feed into planning of next batch	

Figure 20 CoMoUK's guidance on the timeline and stage gates to be followed when setting up a mobility hub network

From Cenex's experiences of working on a mobility hub project for Plymouth, the Phase 1 consultation can take up to 6 months to ensure that all views are heard, and that appropriate time is given to assessing potential locations for a mobility hub. If a potential location is identified that is not on council-owned land, then the owner of the land should be engaged with as soon as possible to understand their potential interest in the project.

Key stakeholders that should be engaged at this stage include:

➜ Architects

➜ MaaS operators

→ Shared mobility providers

→ Mass transit providers

→ Chargepoint providers

→ The general public/residents close to the proposed locations

Once the key elements of the mobility hubs have been agreed and approved – along with their locations, design and size – it is at this stage that a local authority should tender for services (not just vehicles, but the digital aspect as well). It is important at this stage to ensure that a fully functioning and easy to use mobility hub network is provided, and that arrangements for data sharing are in the tender. Only through data sharing/open data can a unified MaaS system be provided where users can switch between modes easily and pay through the same app.

Once the operators have been agreed, along with a MaaS provider (this could also be internal, like Nottingham's Robin Hood Network), a tender for building the mobility hubs should be opened. As part of this process, the local DNO should be involved to understand the potential energy requirements and civil/electrical costs associated with the project.

Would you like to see mobility hubs where you live? If so, think about where these could be located and what size of mobility hub would be best suited to this location.

3.2.1 Data and data sharing

Shared mobility vehicles are able to generate considerable data because they are largely app- and location-based services. Many cities have mandated data sharing as part of a scheme's licence to operate. However, often cities are unsure what to do with the data once they have it, or have data that provide no focus or purpose.

First and foremost, cities need to consider what the data will be used for. This will help to form the questions they have in terms of their capacity to analyse the data, what data they require and to what depth (remembering that operators cannot provide all data for operational, technical or legal reasons, e.g. GDPR), and in what format the data should come.

The data produced typically fall into two main categories: user data and operational data. User data could include gender and age of users, as well as information on what modes were replaced by shared mobility trips. Operational data could include start and end points, date, time and duration. Both data sets improved understanding of travel and transport in a city; however, thought should be given to the end goal of a city, and what data will be most useful to reach their strategic goals.

Operational data is likely to be of greatest use when analysing transport systems and ensuring a network that works in harmony for the greatest benefit to the city. When stored and analysed in time series, operational data can help local authorities to visualise and learn from daily, weekly and monthly shared mobility usage patterns: e.g. when and where traffic is more intense, and parking has more peaks. From this they can identify priority areas for improvement and, coupled with data from public transport and parking ticketing, understand the workings of the transport network as a whole.

Figure 21 Open data connections between the internet and other sources

Open data

In order to make the most of the data produced by shared mobility vehicles, this data should be accessible to other companies.

There are multiple definitions of open data, but for the case of shared mobility the following definition is the most useful, split into three main components.

1. Availability and access: This means providing data as a whole in a convenient and modifiable form. This is key to ensuring that open data is fully open. If the data is provided in a format that requires a very specific program to view it, then it is not open.

2. Re-use and redistribution: In order for data to be considered fully open, you need to have the ability to redistribute and re-use this data. And most importantly this includes intermixing with other data, as that is where the true power in open data lies. This point links nicely back to the first, as a reminder that data should be presented in a convenient format that will allow mixing of datasets.

3. Universal participation: This builds on the first two points and specifies that there should be no discrimination for the use of this data; discrimination could mean providing data in a format that requires a program you have to buy in order to read it, or providing data to some people but not others. In order for data to be considered truly open, there should be no restrictions on the data itself.

The biggest benefit of truly open transport data is the chance for interoperability: that is the ability for various transport systems, operators, users and infrastructure to operate not just alongside each other, but as part of one another. Through this, cities can more easily operate Mobility as a Service (MaaS) platforms. MaaS platforms are a one-stop-shop, where users can purchase single- or multiple-mode travel tickets all on the same app. This allows users to see all the transport modes available to them, where they are, the costs and potential journey times, allowing users to make informed decisions about their travel.

Only through data sharing/open data can a unified MaaS system be provided, where users can easily switch between modes and pay through the same app.

Data sharing and open data are often a controversial topic, so consider the following questions.

What data would you be happy for shared mobility operators to have, and how do you think it would help the wider community? If you are wary of sharing your data, explain why and what policies would need to be in place for you to be willing to share this information.

3.2.2 Real world deployment

Although the theory behind deployment of shared mobility may be great, studying real world examples is even better. Below are a few case studies of how leading cities have introduced shared mobility, mobility hubs and data sharing.

Read through these case studies, then consider whether similar schemes could be possible where you live.

Mobility hub: Leuven, Belgium

Leuven is a mid-sized city with about 100,000 inhabitants, situated 20 km east of Brussels. Because of its location in the slipstream of Brussels, Leuven experiences a high level of traffic congestion. In the last decade, Leuven has implemented a sustainable urban transport strategy, in which the development of shared mobility services and multimodal transport are key elements.

As part of the Interreg EHubs project, Leuven is in the process of developing up to about 50 mobility hubs throughout the city. With an already strong cycling culture and good infrastructure, they are aiming to build on this with a joined-up approach to electrification and shared mobility.

The importance of developing the digital as well as the physical infrastructure for shared mobility is seen as a key priority, and they have worked to develop a MaaS platform that will give easy access to all travellers.

Leuven has already signed contracts with a number of operators to provide shared mobility services, including Blue bikes (non-electric shared bikes), Cargo (cargo bikes), Urbee (e-bikes), and Cambio (car clubs).

In Leuven, there will be larger mobility hubs placed at strategic locations, connected to other modes (e.g. public transport); but also smaller mobility hubs in residential areas.

At the residential level, there was public consultation to ascertain end users' preferences; the results will be integrated in a bottom-up approach during selection procedure. Within the next three years, 50 mobility hubs will be deployed, with the first 15 expected by the end of 2021.

Mobility hub: Plymouth, UK

As part of the Transforming Cities Fund (TCF), Plymouth is aiming to develop a network of mobility hubs around the city, with up to 50 mobility hubs deployed within the next 3 years.

The Mobility Hub Network project meets the aims of the TCF through the provision of low carbon shared transportation and new charging infrastructure, connecting the region's two TCF corridors, increasing connectivity to key employment markets, education, health and leisure facilities, and services. It aligns with the policies of the Department for Transport's Road to Zero Strategy 2018, which specifically detail the crucial role of charging facilities in meeting the Strategy's objectives.

The mobility hubs project will provide at least 300 electric vehicle charging points, 400 e-bikes, car clubs, 0.5 megawatts of solar carports, and a smart booking system. Local residents, employees, businesses and visitors will be able to plan their journeys to use public and shared transportation, both in the city and on the main routes into Devon and Cornwall.

Cenex were commissioned to perform a site selection exercise to determine possible mobility hub locations for Plymouth, and then to assess these locations in accordance with key criteria such as: convenience to the user, ease of installation, visibility and accessibility, ease of modal shift, safety, and practicality of additional facilities such as toilets and cafes.

A shortlist of the best ranked sites was produced, along with a delivery plan for these sites. Plymouth aims to deploy the first of these mobility hubs by 2022, with the project coming to a conclusion by 2023.

Data sharing: Rennes, France

Rennes was one of the first cities in France to launch an open data portal to power a digital ecosystem. A private company, Data2B, compiled an open data platform and implemented a project with the mobility provider Keolis, who operate a local transport network.

Predictive software was developed, aimed at improving the accessibility of local buses. It combined weather, event, historic and live ticketing datasets to predict how full buses would be. The application allows riders to have a better idea of when the next bus is coming, or whether they should wait for a second bus with more space. In addition, Keolis itself benefits in improved operational efficiency.

The service provides more accurate readings on when and where more buses should be deployed in accordance with major events, but can also help the operator decide when to send a smaller bus, helping to reduce fuel consumption and carbon emissions.

Alongside this project, Data2B have also worked with local bike-sharing providers to improve the efficiency of bike redistribution across its network. This helps teams decide where and when they should move bikes between stations.

→ It predicts which stations will have high demand for bikes, or high demand for spaces

→ It creates more efficient routes for teams moving bikes around

→ The network is better optimised for users, and the operator saves time, money and carbon emissions

Can you think of any other good examples of shared mobility being implemented around the world, perhaps where you live or somewhere you visited? What was your experience?

3.3 Part 3 Quiz: Test what you've learned

You have now reached the end of Part 3, in which we've covered city planning, mobility hubs and aspects to do with data. Here is a quick overview of some of the key lessons from this section:

KEY TAKEAWAYS

1. When planning for shared mobility, there are a lot of things to take into consideration, the key elements being: cooperation between local authorities and private operators, reallocation of public space historically allocated to private car use, the location where shared mobility vehicles are placed, infrastructure requirements (parking, bike racks, charge points), the fees that should be charged to operators to conduct their business, safety of users, and social and economic demographics of residents and specific areas.

2. A mobility hub is a recognisable place with an offer of different and connected transport modes, supplemented with enhanced facilities and information features, to both attract and benefit the traveller. A mobility hub is designed and is spatially organised in an optimal way so as to facilitate access to and transport between modes, including human-powered and shared modes, as well as providing extra transport-related and digital services. Spread over an area, mobility hubs provide an unambiguous, recognisable network of defined areas that provide services to connect people through sustainable travel and improving the public realm.

3. In order to make the most of the data produced by shared mobility vehicles, it is important that this data can be accessed by other companies in the form of open data, allowing interoperability. This is the ability for various transport systems, operators, users and infrastructure to operate not just alongside each other but in partnership with one another. Cities can then more easily operate Mobility as a Service (MaaS) platforms: a one-stop-shop where users can purchase single- or multiple-mode travel tickets off the same app. This allows users to see all transport modes available to them, where they are, the costs and potential journey times, allowing users to make more informed decisions about their travel.

Now you've finished Part 3, do try this short multiple-choice quiz to consolidate your knowledge of this section. Then, since you have now covered all the learning content of the book, move on to the final test of the book.

1. Which of the following does NOT need to be taken into consideration when deploying shared mobility?

1. Location of infrastructure

2. The distribution of HGV depots

3. Social and economic demographics of citizens

4. Operator fees charged by cities and local authorities

2. Why are train stations considered a good location for shared mobility?

1. Train stations are always near built-up areas.

2. People who use trains are young and more likely to use shared mobility.

3. People who use trains like sharing.

4. They are a key start and end point for journeys, and can offer a substitute for taxi use.

3. How long does a mobility hub typically take to be developed and built?

1. 2 – 6 months

2. 10 – 15 months

3. 24 – 28 months

4. Over 3 years

4. Which of the following is key to the successful deployment of Mobility as a Service (MaaS)?

1. GDPR rules

2. Cheap development costs

3. Open data and data sharing

4. Memorandum of understanding

5. Data2B's data sharing activity in Rennes, France, optimised the e-bike network. Which of these was not a benefit of implementing their open data platform?

1. Created more efficient routes for teams moving bikes to stations

2. Reduced the likelihood of vandalism

3. Predicted which stations would have high demand for bikes or parking spaces

4. Saved the operator time, money and carbon emissions

You can find the answers to this quiz at the end of the book.

Closing remarks

At Cenex we regularly refer to using 'the right transport for the right journey'. Shared and sustainable mobility solutions are a key step in ensuring that this vision is achievable. While private ownership of vehicles has a long-standing history, it is not a cost-effective form of transport and nor is it an efficient use of resources. If we are to achieve the global net zero ambitions by 2050, then this will require substantial changes to the way that we work, live and move. As we look to move to more sustainable lifestyles, the shift away from ownership to sharing is an important, in fact critical step.

Ten years ago this step would have seemed impossible, but innovations around smaller forms of transport, such as e-scooters and e-bikes, as well as innovations in digital platforms, mean that sustainable and shared transport has become more than just possible – it's become easy! Combined with ever increasing costs for fuel and the cost of living in general, sustainable and shared mobility may even take the leap from possible to desirable.

We hope that you have absorbed some valuable information that will go on to help you in your career and daily life. Do continue exploring the world of sustainable transport and shared mobility!

Find out more...

If you would like to find out more about some of the other areas Cenex is researching, you can join one of our online courses by visiting www.futurelearn.com/partners/cenex, or visit the resources section of our website (www.cenex.co.uk/resources) where we store all our free, public resources, including white papers, technical reports and webinars.

Glossary of terms

Abbreviation	Full Term	Definition
CAV	Connected and Autonomous Vehicles	A vehicle that is capable of driving without human intervention, with technology that enables it to communicate and exchange information wirelessly with other vehicles, infrastructure, other devices outside the vehicle and external networks.
CO_2	Carbon dioxide	A colourless and odourless (greenhouse) gas which is released through human activities such as deforestation and burning fossil fuels, as well as natural processes such as respiration and volcanic eruptions.
EV	Electric Vehicle	Any vehicle powered by an electric drivetrain.
EVSE	Electric Vehicle Supply Equipment	The equipment used to re-charge an EV.
ICE	Internal Combustion Engine	Referring to the internal combustion engine particularly of fossil-fuelled vehicles.
GHG	Greenhouse Gas	Any of various gaseous compounds (such as carbon dioxide or methane) that absorb infrared radiation, trap heat in the atmosphere, and contribute to the greenhouse effect.
MaaS	Mobility as a Service	A single digital transport service platform that enables users to access, pay for and get real-time information on a range of public and private transport options.
NGO	Non-Governmental Organisation	A non-profit organisation that operates independently of any government, typically one whose purpose is to address a social or political issue.
SAV	Shared Autonomous Vehicle	A traditional vehicle converted to provide autonomy via stack technology, or a purpose-built vehicle designed for safely moving people from point A to point B. Different from driverless cars, self-driving cars, or robo-taxis, an SAV is designed for sharing between multiple users.
SMEs	Small and Medium Enterprises	The UK definition of SME is generally a small or medium-sized enterprise with fewer than 250 employees.

SUMP	Sustainable Urban Mobility Plan	A planning concept created and applied by cities and local authorities for strategic mobility planning. Its principal aim is to improve accessibility and provide high-quality, sustainable mobility for the entire urban area.
TCF	Transforming Cities Fund	This was a Department for Transport-backed £2.45 billion capital grant transport fund which aimed at driving up productivity through investments in public and sustainable transport infrastructure in some of England's largest city regions.
V2G	Vehicle-to-Grid	A type of EVSE which allows bi-directional energy flows (i.e. both charging and discharging of energy) between the EV and the device/network it is connected to. This has also become the industry standard term for all bi-directional charging applications.

Quiz answers

Below are the answers to the quizzes from the end of each part, and the final test.

Part 1 An Introduction to Shared and Sustainable Transport

1. Which of these vehicles has the lowest carbon footprint per person, per kilometre travelled?

Answer: Option 2, E-bike

2. Congestion represents wasted time for drivers. What is the average amount of lost time per person per year in the UK due to congestion?

Answer: Option 3, 178 hours

Equivalent to around half an hour per day

3. Which of the following is NOT a safety measure that can be implemented for shared mobility?

Answer: Option 1, The use of green electricity to recharge vehicles

4. What are the two main operating models used for car clubs?

Answer: Option 4, Station-based and free-floating

5. If steady growth of car club members is maintained, how many global users will there be by 2030?

Answer: Option 4, 100 million

Part 2 What is Micromobility?

1. Which of the following is NOT a form of micromobility?

Answer: Option 3, Electric milk float

2. What is the typical range of an electric scooter off a single charge?

Answer: Option 3, 40 miles (64 km)

3. Which of the following does NOT provide a safety hazard for e-scooter users?

Answer: Option 1, Bike lanes. Bike lanes can help to create a safe environment for users.

4. Which of the following is NOT considered a benefit for a docked e-bike system?

Answer: Option 4. Vehicles are in fixed locations aiding door-to-door travel. This is not a benefit, and is a limitation of a docked system.

5. In which of the following areas are e-bikes unlikely to be deployed?

Answer: Option 3, Shopping malls

Part 3 How is this Applied to City Planning and Mobility Hubs?

1. Which of the following does NOT need to be taken into consideration when deploying shared mobility?

Answer: Option 2, The distribution of HGV depots

2. Why are train stations considered a good location for shared mobility?

Answer: Option 4. They are a key start and end point for journeys and shared mobility can replace taxi use.

3. How long does a mobility hub typically take to be developed and built?

Answer: Option 2, 10 – 15 months

4. Which of the following is key to the successful deployment of Mobility as a Service (MaaS)?

Answer: Option 3, Open data and data sharing

5. Data2B's data sharing activity in Rennes, France, optimised the e-bike network. Which of these was NOT a benefit of implementing their open data platform?

Answer: Option 2, Reduced the likelihood of vandalism